WATER POLLUTION: NEW RESEARCH

WATER POLLUTION: NEW RESEARCH

A. R. BURK
EDITOR

Nova Science Publishers, Inc.
New York

For permission to use material from this book please contact us:
Telephone 631-231-7269; Fax 631-231-8175
Web Site: http://www.novapublishers.com

LIBRARY OF CONGRESS CATALOGING-IN-PUBLICATION DATA

Water pollution : new research / A.R. Burk, ed.
p. cm.
Includes index.
ISBN 1-59454-393-3 (hardcover)
1. Water--Pollution. I. Burk, A. R.
TD420.W34253 2005
363.739'4--dc22 2005005549

Published by Nova Science Publishers, Inc. ✦ New York

CONTENTS

PREFACE

Pollution is undesirable state of the natural environment being contaminated with harmful substances as a consequence of human activities so that the environment becomes harmful or unfit for living things; especially applicable to the contamination of soil, water, or the atmosphere by the discharge of harmful substances. In addition to the harm, either present or future and known or unknown, to living beings, pollution cleanup and surveillance are enormous financial drains of the economies of the world. This book gathers leading research from throughout the world dealing with water pollution.

The presence of heavy metals in the sludges produced in wastewater treatment plants restricts their use for agricultural purposes. Chapter 1 looks at four different types of stabilised sludges (aerobic, anaerobic, unstabilised and sludge from a waste stabilisation pond) and analyses the main phases to which the heavy metals are associated. Some chemical and agronomic parameters necessary for characterising a sludge as suitable for use as amendment were determined. The aim of this study is to compare the bioavailability level and the potential mobility of heavy metals in different sewage sludges using simple extraction procedures commonly applied to soils. To know the extraction efficiency, an X-ray diffraction analysis (XRD) has been carried out to the sludge and the residues of simple extractions. This allows a better knowledge of the sludges behaviour once they were added to soil. It was confirmed that the total concentration of heavy metals did not exceed the limits set out by legislation with the exception of Cr in the anaerobic sludge that triplicates the values established in the American legislation. Also, it was verified that the stabilisation treatment undergone by the sludges strongly influences the heavy metals bioavailability level and the phases to which they are associated.

Environmental problems in beet molasses fermentation factories are mainly related with the production of large quantities of a high strength effluent known as vinasse. Chapter 2 reports on the main characteristics of both beet molasses and vinasse and summarizes methods for using and treating molasses processing effluents. Suitable solutions have to be proposed due to the more restrictive legislative rules. Anaerobic digestion is an economically viable and widely used vinasse treatment scheme. Anaerobic digestion is usually followed by an aerobic treatment to meet standard discharge values. However, these treatments are not able to degrade colored compounds in the effluents. Therefore, the anaerobic-aerobically treated effluent has negative environmental impact if released into surface and ground waters due to its brown color. Much of this chapter focused on characterize the colorants present in the molasses processing streams and on examining those treatments which are expected to degrade the non-biodegradable colored compounds. Among them, chemical oxidation with ozone is a promising technique for removing color from both the vinasse and the anaerobic-aerobically treated effluent without producing by-products that need further treatments.

The loss of phosphorus (P) in drainage waters can cause eutrophication and impairment of surface water quality. Dissolved organic P (DOP) is available to algae and also leaches farther down the soil profile than dissolved inorganic P, especially in soils that have received long-term applications of manure or effluent. Best management stipulates that P fertiliser applications should account for P added in effluent to not overload soil with P and risk P loss to surface waters. Hence, the aim of chapter 3 was to compare the form and losses of P in mole-tile drainage waters from a plot receiving dairy effluent to one that received P at the same rate but as superphosphate. Data over 3 years indicated that much more P was lost from the effluent-irrigated plot (mean TP load = 0.41 kg ha^{-1}) compared to the non-effluent plot (mean TP load = 0.20 kg ha^{-1}), especially via incidental transfers (events coinciding with effluent application or within a week of cattle grazing, up to 56% in the first year). Losses of DOP were, on average, 3.7 times greater in the effluent irrigated compared to the non-effluent plot. When incidental transfers were excluded, a significant relationship still occurred between the difference in DOP concentrations between plots, and event number, since irrigation began, indicating that DOP loss is enhanced by short-term effluent applications. Analysis of soil samples taken from each plot indicated that more P was held in the Ca-P (due to Ca application in effluent), water soluble P and NaOH organic P fractions of the effluent irrigated soil than the non-effluent soil. Analysis of raw effluent and effluent and non-effluent treated soils by ^{31}P nuclear magnetic resonance spectroscopy indicated that much organic P (present in the NaOH organic P fraction) was phytate (450 and 386 mg kg^{-1} in the 0-2 cm depth of effluent and non-effluent treated soils, respectively). The enhanced loss of P in drainage waters from effluent irrigated soil was attributed to an increase in soil pH, and competition of P sorption sites by phytate, organic compounds and P from P-saturated colloids in the effluent. Since DOP is poorly held by soil compared to orthophosphate, we concluded that any mechanisms influencing P desorption will influence DOP loss more than DRP, especially in terms of P desorption. This highlights the need to account for DOP loss and bioavailability in soils receiving effluent even in the short-term.

In recent years invertebrates are gaining importance for bioindication due to their integration of pollutants in space and time, the short life cycle and the relative ease of collection. Nevertheless, it is necessary to reflect upon the methodology of collecting and sample preparation as previous metal studies on invertebrates, living in urban environments, suggest that a variety of factors (e.g. trophic level, strategies for detoxification, defecation, and type of trapping) are responsible for differences in sample concentrations. In order for the authors of chapte 4 to obtain basic information for designing standardised test preparation methods, the heavy metals Zn, Cu, Cd and Pb were measured in gastropods (*Xerolenta obvia*), oligochaetes (*Lumbricus terrestris*), isopods (*Armadillidium vulgare* and *Trachelipus rathkei*) and carabids (*Harpalus rubripes* and *Calathus fuscipes*) using different sampling methods and different modes of sample treatment. Isopods and carabids trapped with formalin-pitfalls were compared to manually collected specimens (which were allowed to defecate). Additionally effects of defecation were studied by feeding cellulose as an accelerator of gut passage in gastropods and oligochaetes. Effects of sample cleaning were determined in isopods (*A. vulgare*) between deionate-washed and ultrasonic-cleaned individuals. Type of sample drying is very important as volatile elements like Cd, Hg, and Se may exhaust during sample preparation. For that reason Cd concentration was compared between heat-dried versus freeze-dried isopods. Sex-specific differences in metal concentrations were investigated in carabids. Based on the author's and previous results, invertebrates should be: collected *in vivo*, allowed to defecate, be freeze-fixed and (at least arthropods) ultrasonic-cleaned, determined to species level and in certain groups (carabids) also to sex, and then be sized or sorted by size (and age) before further preparation and analysis. If any of these treatments is impractical, comparable sampling and preparation methods are recommended as a minimum requirement in order to avoid bias in the results and/or interpretation.

Advanced oxidation processes (AOPs) are emerging and promising technology both as an alternative treatment to conventional wastewater treatment methods and enhancement of current biological treatment methods especially dealing with highly toxic and low biodegradable wastes. In chapter 5, the various advanced oxidation processes and their common principles were briefly discussed. In addition, reactors and light sources used most frequently are given in summary. Finally, the application of AOPs to wastewater treatment focusing on some case studies (coloured textile effluent, olive mill effluent, domestic wastewater) is presented.

The adsorption of [1-(6-chloro-*3*-pyridinylmethyl)-*N*-nitro-2-imidazolidin-mine] (imidacloprid) on heat treated kerolite samples at 110°C (K-110), 200°C (K-200), 400°C (K-400) and 600°C (K-600) from pure water solution at 25°C has been studied in chapter 6. In addition, the influence of the presence of 0.1 M KCl in the medium and the influence of temperature, by performing the experiments at 10°C and at 40°C, was also investigated for the sample K-600. The evolution of surface properties of kerolite samples such as specific surface area and porosity after heat treatment were analysed. The clays were characterised by using: FTIR spectroscopy, XRD diffraction, TG and DTG analysis, surface analysis and Hg porosimetry. The experimental adsorption data points were fitted to the Freundlich equation in order to calculate the adsorption capacities (K_f) of the samples; K_f values range from 242 mg kg^{-1} for the K-110 sample up to 1005 mg kg^{-1} for the K-600 sample, that this, the stronger heat treatment, the more effective the adsorption of imidacloprid from pure water. The presence of 0.1 M KCl in the medium tends to increase slightly the adsorption of imidacloprid in the range of experimental conditions used. The variation of temperature did not show a clear influence on adsorption of imidacloprid from both, pure water and 0.1 M KCl solutions, since the adsorbed amount of imidacloprid seems to be dependent from the imidacloprid initial solution concentration (C_0). Nevertheless, the adsorbed amounts of imidacloprid at 25°C and 40°C are, in general, higher than those corresponding to 10°C. The results obtained from this work could be of interest by showing the better experimental conditions for the potential use of heat-treated kerolites -especially that activated at 600°C- for adsorption of imidacloprid. It seems, in general, that by conducting the adsorption process at 25°C in presence of 0.1 M KCl solution, the optimum adsorption of imidacloprid could be achieved, independently of the initial solution concentration of imidacloprid.

A ready biodegradability test system is established which is based on the quantification of oxygen and carbon dioxide in the head space of batch vessels using gaschromatography with thermal conductivity detector. The test system is in accordance with a standard procedures ISO 10708 (BODIS-Test), ISO 11266 and ISO 14593 (ISO CO_2 Headspace Test) with slight modifications. It is applicable for the assessment of ready biodegradability in water and soil. In this study the "O_2/CO_2 Headspace Test with GC-TCD" is used for the assessment of synthetic ester lubricants. The effects on biodegradability in water and soil of both additives and usage are examined and discussed in chapter 7. Ester based cutting fluids, gear fluids and hydraulic fluids with and without additives are used under defined conditions in machine tools and hydraulic and plain bearing test benches. The lubricants are characterised additionally with respect to kinematic viscosity, acidity and elemental composition. Furthermore, a formulated mineral oil is characterised before and after usage in the hydraulic test bench. Early results clearly show that the mineral oil is far less biodegradable than the ester oils and that their biodegradability is not affected by usage. The biodegradability of the ester oils mainly depends on the characteristics of the base fluids and is not affected by the additives. Antioxidants influence stability and biodegradability indirectly, since they prevent oxopolymerisation effects. Other effects of usage on biodegradation of lubricants are not detected. In this context, the antioxidants ensure ready biodegradability and have a positive effect on the environmental fate of synthetic ester lubricants. These statements can be affirmed in advanced studies with the results on biodegradability in soil. Furthermore, the fluids show a lower biodegradability in soil within 32 days than in water.

Halogenated organic compounds in water can react under anaerobic conditions with zero-valent iron Fe^0. While halogenated compounds are reduced via intermediates to non-toxic simple hydrocarbons, Fe^0 is oxidized to Fe^{2+}. An excess of hydrogen gas is also produced by the reaction of iron with water. Dehalogenation reaction can be utilized for decontamination of underground water in passive permeable reactive barriers, where water flows spontaneously through a layer of granular iron. The basic parameters of the dehalogenation reaction and field application of the reaction including applications of isotopes and geochemical modeling are discussed in chapter 8.

There is increasing experimental evidence that protists can be exploited as excellent tools for environmental biomonitoring, either as bioindicators of pollution or bioassays to evaluate the effects of toxic compounds. Protists exhibit the fundamental features required to perform rapid biotests with a high degree of reproducibility. Due to their nature as single-cell eukaryotic organisms, they are able to respond directly to environmental stimuli, and multiply through short cell cycles. Chapter 9 explains the extensive efforts that are currently devoted to enlarging the panel of organismic systems sensitive to environmental perturbations, such as the occurrence of neurotoxic drugs, which are largely employed on agricultural sites for their insecticide properties. It has long been known that the first target of such toxic compounds is the inhibition of cholinesterase (ChE) enzyme activities, which in turn represents their specific biomarker. In this context, the identification of ChE activities both in the freshwater ciliate, *Paramecium primaurelia,* and the slime mould *Dictyostelium discoideum*, appears interesting, as the sensitivity of these enzyme activities to the presence of neurotoxic drugs allows us to develop a biomonitoring system that exploits these single-cell eukaryotes for the pre-chemical screening of freshwater and moist environments. The ChE activities have been assumed to be involved in cell-cell interactions taking place in *Paramecium* conjugation, as well as in *Dictyostelium* cell migration and aggregation. In fact, the ChE activity inhibition caused by exposure to neurotoxic drugs has been found to perturb mating cell pairing in *P. primaurelia* and cell streaming in *D. discoideum*. The alteration to these essential processes for the development in the respective life cycles could result in important ecological effects on natural microbial communities.

In: Water Pollution: New Research
Editor: A. R. Burk, pp. 1-27
ISBN: 1-59454-393-3

Chapter 1

AVAILABILITY AND POTENTIAL MOBILITY OF HEAVY METALS IN SLUDGES USING SIMPLE EXTRACTION PROCEDURES

A. Fuentes, M. Lloréns,[1] J. Sáez, M. I. Aguilar, J. F. Ortuño and V. F. Meseguer
Department of Chemical Engineering. University of Murcia. Espinardo. 30071 Murcia. Spain

ABSTRACT

The presence of heavy metals in the sludges produced in wastewater treatment plants restricts their use for agricultural purposes. This study looks at four different types of stabilised sludges (aerobic, anaerobic, unstabilised and sludge from a waste stabilisation pond) and analyses the main phases to which the heavy metals are associated. Some chemical and agronomic parameters necessary for characterising a sludge as suitable for use as amendment were determined. The aim of this study is to compare the bioavailability level and the potential mobility of heavy metals in different sewage sludges using simple extraction procedures commonly applied to soils. To know the extraction efficiency, an X-ray diffraction analysis (XRD) has been carried out to the sludge and the residues of simple extractions. This allows a better knowledge of the sludges behaviour once they were added to soil. It was confirmed that the total concentration of heavy metals did not exceed the limits set out by legislation with the exception of Cr in the anaerobic sludge that triplicates the values established in the American legislation. Also, it was verified that the stabilisation treatment undergone by the sludges strongly influences the heavy metals bioavailability level and the phases to which they are associated.

[1] Department of Chemical Engineering. University of Murcia. Espinardo. 30071 Murcia. Spain. E-mail: llorens@um.es

INTRODUCTION

The accumulation of sewage sludges from urban wastewater treatment is a growing environmental problem and it has been calculated that 60 g of dry solids per person and day are produced by those plants which incorporate a secondary treatment step (Lee et al., 2002; Balmer, 2001). Between 1992 and 2000 the production of sludge in the EU reached around eight million tons of dry waste matter (Magoarou, 2000), while a similar amount was produced in the USA in 2000 (Englande and Reimers, 2001). Following the construction of a large number of new treatment plants in China, it is expected that about four million tons of dry matter will be produced in coming years (Lee et al., 2002).

The application of such sludges as fertilisers or organic regenerators of the soil presents itself as an attractive possibility because this will enable valuable components (organic matter, N, P and other nutrients necessary for plant growth) to be recycled (Hernández, 1991). However, this practice represents a potential threat to the environment because of the possible high heavy metal content. This problem may be aggravated if the toxic metals are mobilised in the soil and taken up by plants or transported in drainage waters (Angelidis and Gibbs, 1991; Mingot et al., 1995; Zufiaurre et al., 1998; Álvarez et al., 2002).

After addition to the soil, the heavy metals contained in sewage sludges may remain dissolved in the soil solution or be fixed by processes of adsorption, complexation or precipitation. They may also be taken up by plants and thus enter the food chain, pass to the atmosphere through volatilisation or leach into surface and underground waters, where they can remain in solution or be precipitated into the sediment, from where they may enter the food chain again (Hooda and Alloway, 1994; Moreno, 1997).

Table 1 illustrates the principal supports which can fix the metals and the physical or chemical processes responsible for this fixation.

Table 1. Principal supports capable of fixing metals

Sediment or sludge phase	Processes responsible for fixation
Vegetable remains (mainly silicates)	Metal strongly complexed in inert form
Metals bound to carbonates, sulphides and hydroxides	Precipitation by overcoming solubility rate
Clay minerals	Physical adsorption (electrical attraction) Chemical adsorption by exchange of H of SiOH, AlOH and Al $(OH)_2$
Lipids, humic substances, organic wastes	Physical adsorption Chemical adsorption (COOH y OH) Complexes
Iron and manganese oxides and hydroxides	Physical adsorption Chemical adsorption Co-precipitation by saturation
Calcium carbonate	Physical adsorption Pseudomorphism Co-precipitation

The different chemical forms taken by metals in sludges and the processes of transformation to which they are submitted after addition to the soil influence their mobility and availability. The following fractions can be distinguished (Carre and Welte, 1986; Calvet et al, 1990):

1. Water soluble.
   - free ions.
   - inorganic complexes.
   - organic complexes.
2. Exchangeable, that is, bound to the exchange complex.
3. Precipitated as inorganic compounds.
4. Complexed in humic substances of high molecular weight.
5. Adsorbed or occluded in oxides and hydroxides.
6. Precipitated and insoluble as sulphides.
7. Retained in the crystalline structure of primary minerals.

The water-soluble fraction is the most mobile fraction and the most readily available to plants, although the exchange fraction may also be released to the soil solution and become available (Gambrell, 1994). Quite the opposite is the case with the metal fraction bound to the crystalline structure of the mineral components of the sludges. This fraction is not immediately available and will only become so by meteorisation of these minerals.

To evaluate the environmental impact of these metals, it is not sufficient to determine the total content since their behaviour in a given medium and their capacity for mobilisation are equally important and these factors will depend on the chemical form in the sludge (Legret, 1993; Zufiaurre et al., 1998; Pérez-Cid et al., 1999; Fernández et al., 2000; Fuentes et al., 2004a; Fuentes et al., 2004b).

Chemical speciation can be defined as the process of identification and quantification of different species, forms or phases present in a material. These species can be defined (1) functionally, for example those species that can be assimilated by plants, (2) operationally, according to the procedures or reagents used in their extraction and (3) specifically, as particular components or oxidation states of an element. Operationally, the definition the speciation includes simple and sequential extraction to relate the species associated with particular phases of the sludges (Davidson et al., 1994).

METHODS

Sludges

All the sludges studied come from urban wastewater treatment plants located in the Region of Murcia (SE Spain). They were stabilised (or not) in different ways: anaerobically, aerobically, in a waste stabilisation pond (WSP) and one unstabilised.

Equipment

A Perkin Elmer AA300 atomic absorption spectrometer with deuterium background corrector was used to determine the total content of Cd, Cr, Cu, Fe, Ni, Pb and Zn. A fuel-lean air-acetylene flame was used for all the analytes except Cr, which requires the stoichiometric conditions of flame gases.

When the sensitivity of FAAS was insufficient to determine Cd, Cu, Ni and Pb in the extracts, electrothermal atomisation AAS (Perkin Elmer HGA 800 with AS72 autosampler) was used. In the case of Cr the standard additions method had to be used due to interference on the part of the matrix. $NH_4H_2PO_4$ (1000 mg l^{-1}) was used as matrix modifier for determining Cd.

Reagents

All the reagents used were of analytical grade. Deionised water was obtained from a Millipore Milli Q system. Certified aqueous standards of the elements (Panreac, Barcelona, Spain) were used for AAS. All the standards, reagent solution and samples were stored in polyethylene containers previously cleaned with 4M HNO_3 and rinsed with deionised water.

Sample Pretreatment

The sludges from the different stabilisation processes were frozen immediately after collecting (Berrow and Stein, 1983; Welte and Montiel, 1985) and then lyophilised in an Edwards 4K freeze dryer. The sample was then ground and sieved to obtain a fraction of less than 150 μm. The resulting material was stored in polyethylene containers at room temperature until analysis.

Characterisation of the Sludges

Four replicates of each lyophilised and sieved sample were characterised according to different agronomic parameters: pH, electrical conductivity, moisture content, organic matter content, total nitrogen and total phosphorus. These parameters were determined by standard analytical methods, except in the case of nitrogen, which was determined by elemental analysis. The quantity of the various macroelements was ascertained by atomic absorption in the case of calcium and magnesium and atomic emission in the case of sodium and potassium.

Heavy Metal Concentration

The pseudo-total concentration of the metals was determined by FAAS after digestion of the samples with aqua regia (ISO 11466).This International Standard has been proposed for the determination of extractable metals in soil and similar materials (Welte and Montiel,

1985; Scancar et al., 2000; Pueyo et al., 2001). For this, 5 g of sludge samples were weighed into a 500 ml round-bottomed flask and 7.5 ml conc. HCl and 2.5 ml conc. HNO_3 per gram of dry matter were added. The attack was allowed to proceed for 16 hours at room temperature and then for two hours in open reflux condition at 180° C. After digestion, the solution was filtered into a 100 ml flask and the filter paper and residue were rinsed several times with several millilitres of warm (50° C) 2M nitric acid, making the solution up to the mark with the same acid solution once cold.

Extraction of Heavy Metals

Simple extraction permits the mobility and availability of the metals present in sludges to be evaluated, reproducing different conditions in which the sludges may be found after addition to the soil. Figure 1 shows in schematic form the procedure followed for carrying out the simple extractions.

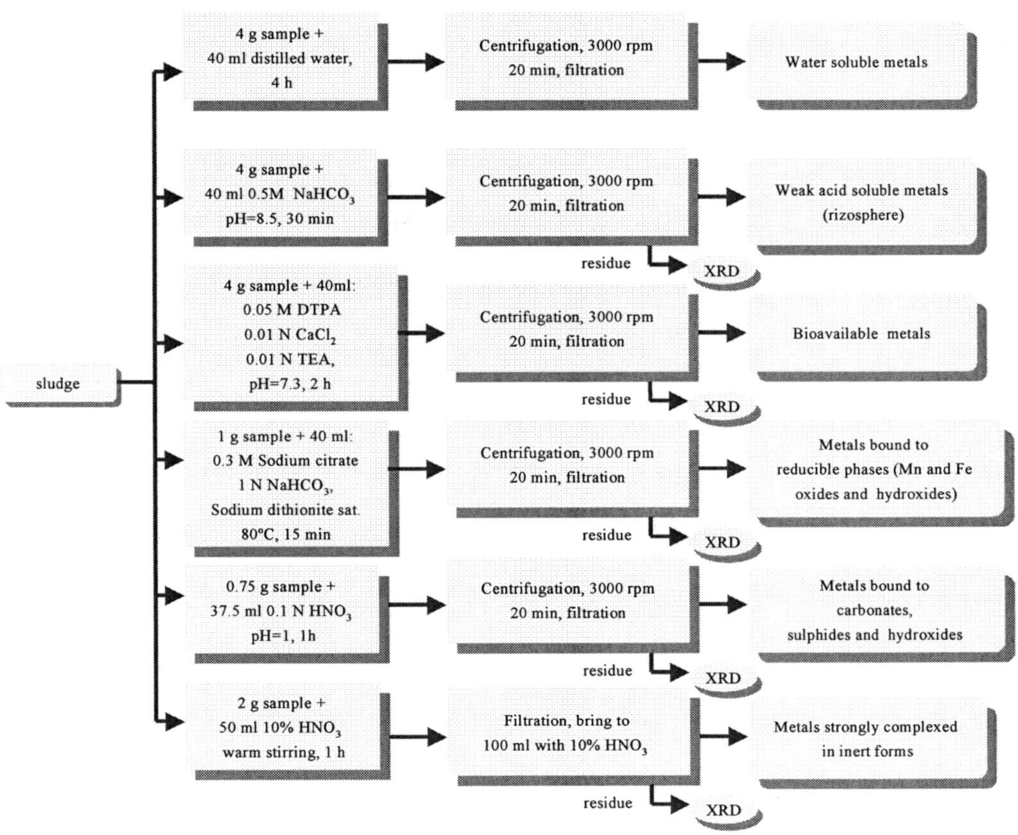

Figure 1. Flow diagram representing simple extraction procedure.

The metals studied were Cd, Cr, Cu, Fe, Ni, Pb and Zn. The extraction media used are described below.

Extraction with Water

It is important to know the mobility of heavy metals in water, especially in the case of sludges destined for application to the soil since it is the water-soluble fraction that is the most mobile and therefore most likely to reach subterranean and/or superficial waters, at the same time representing the most plant-available fraction. However, in general, sewage sludges have a very low water-soluble heavy metal content since the bulk of this fraction is eliminated with the treated waters (Bradford et al., 1971; Moreno, 1985; Campanella et al., 1987; Wong et al, 2001; Fuentes et al, 2004a; Fuentes et al., 2004b).

Reagents

- Distilled water

Procedure

To four grams of sample, add 40 ml distilled water and stir for four hours. Then, centrifuge at 3000 rpm for 20 minutes and separate the supernatant from the residue by decantation. The resulting liquid can be kept in plastic containers until analysis by atomic absorption spectrometry.

Extraction with Bicarbonate

The method used is a modification of the method described for determining available phosphorus by Olsen et al (1954). Schoenau and Karamanos (1993) successfully applied this method for determining the quantity of available P, K and Na.

During extraction, the bicarbonate acts by an ionic effect and through its pH, solubilising phosphates and labile solid compounds of phosphorus, such as phosphates adsorbed in active calcium carbonate, poorly soluble calcium phosphate precipitates and phosphates that are weakly adsorbed to the surfaces of iron and aluminium oxides. Phosphorus is extracted from the soil at a pH of 8.5 (Olsen and Sommers, 1982; Schoenau and Karamanos, 1993).

This method is valid for extracting the heavy metal fraction that is mobilisable in weak acid medium, simulating the conditions that occur in the rhizosphere (García-Rizo et al., 1999; Vidal, 2002).

Reagents

- 0.5 M solution of bicarbonate

Procedure

To four grams of sample, add 40 ml bicarbonate and stir gently for 30 minutes. Separate the supernatant after centrifuging at 3000 rpm for 20 minutes. The metals are then determined by atomic absorption spectrometry (AAS) (Olsen et al., 1954; Olsen et al., 1965; Olsen and Sommers, 1982).

Extraction with DTPA

This method was originally developed by Lindsay and Norvell (1978) to detect deficiencies in metallic micronutrients, plant-available Fe, Mn, Cu and Zn in calcareous soils near neutrality, although it is also considered suitable for quantifying the labile or bioavailable fraction of metals (Petruzzelli, 1989; Calvet et al., 1990; Soon and Abboud, 1993; Wong et al., 2001; Su and Wong, 2003; Fuentes et al. 2004a; Fuentes et al, 2004b).

This extractant was selected from a large number of chelating agents because it is better at complexing Fe, Mn, Cu and Zn simultaneously. Its strong pH dependency can be solved by adding triethanolamine (Liang and Karamanos; 1993). Soon and Abboud (1993) also used this method for determining the quantity of Cd, Ni and Pb that is available.

Reagents

Extractant solution:

- 0.05 M in DTPA (diethylenetriaminepentaacetic acid)
- 0.01N in $CaCl_2$
- 0.1N in TEA, used as pH regulator to attain pH of 7.3.

Procedure

Add 40 ml extractant solution to 4 g of sludge and stir for two hours. Centrifuge for 20 minutes and separate the supernatant from the residue by decanting. Keep the liquid in plastic containers until analysis by AAS (Lindsay and Norvell, 1978; Lindsay and Norvell, 1969).

Extraction in Citrate-Dithionite Medium

The amorphous oxides and hydroxides of iron and manganese play an important role in retaining heavy metals by adsorbing them to the particle surfaces. They are known to have a high capacity to fix heavy metals and to immobilise them in oxidising conditions (Gambrell, 1994, Chlopecka and Adriano, 1996). Furthermore, these compounds are finely disseminated and therefore very active. Sludges with a high iron and manganese oxide content have a great capacity to adsorb divalent metals such as Cu, Mo, Ni, V, Pb and Zn.

The citrate-dithionite medium is especially useful for determining the "free" iron in sewage sludges. Finely divided hematites, goetite, lepidocrocite and ferrihydrite are solubilised, as are non-crystalline iron oxides and organic complexes of iron and aluminium (Mehra and Jackson, 1960), although the medium is less effective for inorganic forms of non-crystalline aluminium (Vidal, 2002).

The same method hardly extracts iron and aluminium from the most saline forms of silicated minerals, which makes it suitable for evaluating non-silicated free iron (Welté et al., 1983; Ross and Wang, 1993; García-Rizo et al., 1999).

Reagents

- 0.3 M dihydrate sodium citrate
- 1N sodium bicarbonate
- Saturated solution of sodium dithionite

The sodium citrate serves as a chelating agent for the ferrous and ferric forms of free iron, while the sodium bicarbonate acts as buffer and the sodium dithionite as reductant.

Procedure

The sample (1g) is placed in a centrifuge tube before adding 40 ml of 0.3 M sodium citrate and 5 ml of 1N $NaHCO_3$. The suspension is then heated in a water bath at 80°C. One gram of sodium dithionite is added and the suspension stirred occasionally for 15 minutes. Care must be taken to avoid heating to more than 80°C to prevent the formation of FeS. After centrifuging for 25 minutes at 3000 rpm, the supernatant is kept in a plastic container until the heavy metals are determined by AAS (Mehra and Jackson, 1960).

Extraction with 0.1N HNO_3 at pH 1

The method used for the extraction with 0.1 N nitric acid, bringing the pH to 1, is a modification of the method involving 1 M nitric acid used to extract the metal fractions co-precipitated with hydrate, carbonate and sulphide oxides or those fractions adsorbed by these species (Soon and Abboud, 1993). To ensure complete solubilisation of the carbonates concentrated nitric acid is added to reach pH 1.

Reagents

- 0.1N nitric acid solution
- Concentrated nitric acid

Procedure

To 0.75 g of sample, add 37.5 ml of 0.1 N nitric acid, adjusting the pH to 1 with a few drops of concentrated acid if necessary. Stir the suspension for one hour and separate the supernatant after centrifuging at 3000 rpm for 20 minutes. After filtering, keep the liquid in a plastic container until determining the metals by AAS.

Extraction with 10% HNO_3

This extraction method indicates the mobility of metals in strong acid conditions, involving a substantial destructive action on the organic matter, and of mineral phases susceptible to solubilisation in acid medium (sulphides, carbonates, phosphates, ...).

Reagents

- Nitric acid solution at 10%

Procedure

Add 50 ml 10% HNO_3 solution to 2 g of sample, warm and stir for 1 hour. After cooling, filter through Whatman n° 2 filter paper. Bring to 100 ml with 10% nitric acid and quantify the metals by AAS.

X-Ray Diffraction

To check the specificity of the reagents used, the residues obtained after each treatment are analysed by X-ray diffraction (XRD), evaluating the variation in mineralogical composition of the residue and untreated sample. This procedure, denominated Differential X-Ray Diffraction (DXRD), shows the effect of the different treatments and is qualitative and only semi-quantitative. It also shows limitations in the determination of the amorphous phases and phases of low crystallinity (Bish, 1994; García-Rizo et al., 1999), so that the way in which these phases are affected by the different extractants cannot be precisely determined precisely (Vidal, 2002).

Procedure

The analysis was carried out in a Philips PW3040 x-ray diffractometer with window split width n° 1, proportional counter and nickel filter. The instrumental conditions for the XRD scan were adapted by varying the sensitivity of the recorder according to the sample (see Table 2).

Table 2. Instrumental conditions used for XRD analysis

PARAMETERS	
Radiation (nm)	1.54
Voltage (KV)	45.0
Current (MA)	40.0
Angular range (° 2θ)	3.02-69.98
Time constant (s)	1

To study the mineralogy, the sample was ground in a ball-mill to less than 50 μm and placed on an aluminium. The qualitative analysis of the treated and untreated samples was carried out using the X'Pert Graphics and Identify program, the most probable phases present being selected in accordance with biographical references found for soils, sludges and sediments (Sirviera and Sommers, 1977; Levi-Minzi et al., 1981; Essington and Mattigod, 1991; Ortuño, 1993).

Results and Discussion

Characterisation of the Sludges

Table 3 shows the main chemical and agronomic parameters obtained during the characterisation of the four types of sludge studied. In general terms, the materials have a pH between 6 and 8, an electrical conductivity slightly below that of manures and are suitable for addition to soils. They have an ash content which varies in relation with their degree of mineralisation, although quite high, as corresponds to materials with a high proportion of silicates in their mineral composition.

The organic fraction of the sludges studied is of note, although the stabilisation process to which they are submitted plays an important role as regards the quantity and quality of the same (Fuentes et al, 2004b). Their total nitrogen content, which varies between 1.7% in the WSP sludge and 4.56% in the unstabilised sludge, gives them an intermediate value as fertiliser. Their C/N values of around 7 are suitable for re-establishing a soil's fertility (Moreno, 1985). As can be seen, the organic matter and nitrogen content diminishes in all the sludges as the degree of mineralisation increases.

As regards the phosphorus content, this may be considered high, and it can be seen that in all the sludges studied the C/P ratio is always<100,_so that the phosphorus occurs in chemical forms that are effective in the medium term in all cases.

The macronutrients studied show the same trend in all the sludges, calcium being the most abundant, followed by magnesium, potassium and sodium. The values determined are similar to those found in the bibliography for sludges, some of which are used for agricultural purposes (Campanella et al., 1987; García et al, 1991; Zufiaurre et al, 1998; Álvarez et al, 2002; Fuentes et al, 2004b).

Table 3. Chemical and agronomic characteristics of the four sludge types studied

	Aerobic	Anaerobic	Non-stabilised	WSP sludge
pH	6.6± 0.1	7.9±0.3	6.0±0.1	8.2±0.1
EC (mS/cm)	2.27±0.06	2.81±0.04	2.3±0.1	1.25±0.04
Organic Matter (%)	64±1	43.5±0.8	65±1	26.3±0.7
Ash (%)	27.8±0.5	44±1	25.6±0.4	50±2
Carbon (%)	41.1	27.06	38.04	19.8
Sulphur (%)	0.71	1.61	0.91	1.05
Nitrogen (%)	4.40	3.27	4.56	1.63
C/N	6.91	7.03	6.89	8.10
Phosphorus (%)	0.77±0.04	1.67±0.04	1.1±0.2	0.42±0.02
C/P	38.6	13.6	27	32.1
Calcium (%)	4.2±0.3	6.77±0.04	4.1±0.1	17±1
Potassium (%)	0.37±0.01	0.28±0.005	0.35±0.002	0.33±0.003
Magnesium (%)	0.54±0.01	0.95±0.06	0.6±0.02	0.89±0.02
Sodium (%)	0.19±0.01	0.15±0.008	0.17±0.009	0.32±0.01

The results represent the mean ± standard deviation for n=4, except in the case of the elemental analysis

Pseudo-Total Content

The use of sewage sludges as fertiliser involves certain restrictions based on their heavy metal content. Table 4 shows the restrictions imposed for the heavy metal content by the EU (86/278/EEC Directive) and the USA (40 CFR Part 503 Sludge Rule).

Table 4. Maximum authorised limits for heavy metals

mg/kg	UE Directive	USA
Cadmium	20-40	39
Copper	1000-1750	1500
Nickel	300-400	420
Lead	750-1200	300
Zinc	2500-4000	2800
Mercury	16-25	17
Chrome	n.s.	1200
Selenium	n.s.	36
Molibdenum	n.s.	18
Arsenic	n.s.	41

Source: Smith, 1996; n.s.: not specified

If the quantity of heavy metals determined in the four types of sludge studied (Table 5) are compared with the maximum permitted values (Table 4), it can be seen that all the values are below the maximum permitted concentrations. However, the quantity of chrome found in the anaerobic sludge was very high and, although the EU directive sets no specific limit for this metal, the American norm sets a limit value of 1200 mg/kg. Proposals have been submitted to the EU (Spinosa, 2001) contemplating a maximum concentration of 1000 mg/kg, meaning that European legislation, to, would consider the sludge from the anaerobic treatment too contaminated for application to a soil. (The high Cr content of the sludge in question was due to it being collected from a wastewater treatment plant that receives wastes from nearby tanning industries). The cadmium content of the same sludge was also very close to the lower limit established by European legislation.

Table 5. Pseudo-total content of heavy metals in each sludge studied

	Aerobic	Anaerobic	Non-stabilised	WSP sludge
Cadmium	1.10±0.07	18.3±0.5	1.14±0.06	11.4±0.4
Chrome	38±3	3809±70	61±1	71±5
Copper	204±5	337±10	146±7	167±7
Iron	5583±331	25208±545	5475±75	8916±45
Nickel	17±2	29±2	25±1	15±1
Lead	58±1	167±3	86±1	250±10
Zinc	487±28	871±37	458±11	697±23

The results are expressed as mean ± standard deviation in mg/kg d.m. for n=4 (with three replicas per sample)

Simple Extraction Applied to Aerobic Sludge

Figure 2 depicts the results obtained after application of the different simple extractions to the aerobic sludge. No metal shows a level that exceeds present day legislation referring to the possible agricultural use of the sludge. Furthermore, the total content of each metal is far from reaching the established legal limits.

Copper mobility is low. Indeed, the proportion present in the most available fractions (water soluble, mobilisable in weak acid and DTPA extractable) is less than 15% of its total content. The quantity retained by the iron and manganese oxides and hydroxides is not significant since that extracted in a reducing medium hardly reaches 5%. A substantial proportion of this metal is strongly associated to the crystalline structure of the sludge, as can be deduced from the low degree of extractability of the copper in nitric acid. Moreover, the sulphide-bound proportion of this element is theoretically low, given the low mobility seen in 0.1 N nitric acid, while the fraction bound to the organic matter is high, as judged from the great difference between the percentages extracted in 0.1 N and 10% nitric acid.

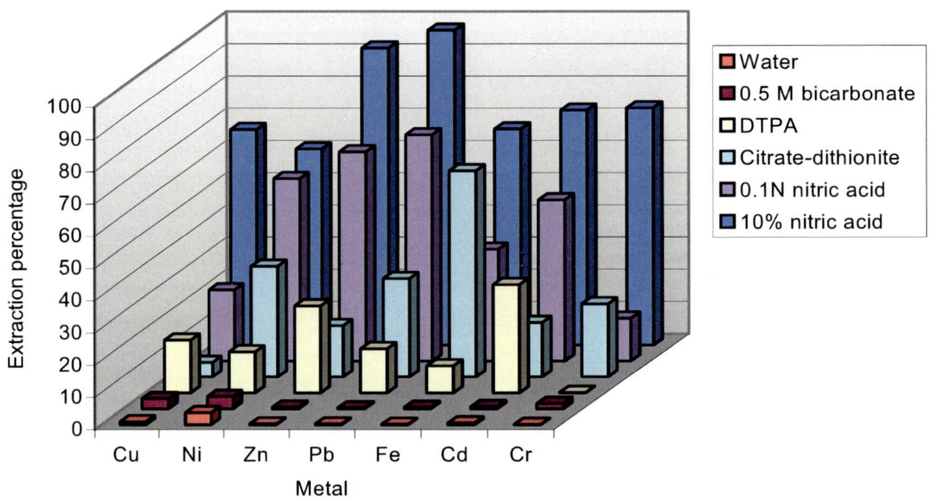

Figure 2. Percentages of each metal extracted by the simple extractions practised in the aerobic sludge

As regards the nickel content, its bioavailability is also low, although the proportion of water-soluble nickel is higher then the corresponding proportion of the other metals. However, the proportion extractable in water, bicarbonate and DTPA does not reach 15%, underlining the low mobility of this element in the aerobic sludge. The proportion mobilisable in chelating-reducing conditions exceeds 30%. The low extraction percentage in a nitric medium, including in strong conditions, shows that a substantial fraction of nickel is associated to the mineral phase of the sludge most difficult to attack, such as the silicates in their different forms (feldspars, phylosilicates, pyroxene, etc.).

Although zinc is among the metals of higher concentration found in the aerobic sludge (around 500 mg/kg) it is hardly soluble in water and bicarbonate. The bioavailable fraction of zinc in this sludge represented about 20% of the total, while that bound to iron and manganese oxides and hydroxides represented about 10%. However, the mobility of this element in nitric medium is high, suggesting that a substantial part of it is bound to the sulphides and organic matter.

In the aerobic sludge, both lead and iron show a low degree of mobility in the most bioavailable fractions, although both are highly mobilisable in chelating-reducing conditions. This indicates that, in the case of iron, 63% of the total is found in non-silicated forms, while around 30% of the lead is associated to the iron and manganese oxides and hydroxides in the sludge. The recovery of lead is practically total in strong acid conditions, meaning that the fraction bound to the crystalline structure of the sludge is very low. Therefore this element is mainly forming sulphides or bound to the organic matter. As regards the iron, its mobility in acid medium is low since a substantial proportion is presumably associated to silicates that are difficult to attack.

Although the solubility of cadmium in water and bicarbonate is low, as can be seen from Figure 2, the percentage of this metal bioavailable to plants can be considered high (around 35%). However, given its low concentration in the aerobic sludge, this does not present any great risk. The quantity of this metal retained by iron and manganese oxides is low and only 70% of the cadmium is mobilised in strong acid conditions, while a substantial part is associated to the silicated forms of the sludge.

Finally, chrome shows a low degree of extractability in all the media assayed. Even in 0.1 N nitric acid, the proportion extracted does not exceed 15%, demonstrating the reduced amount of this metal bound to sulphides. In strong acid conditions, 70% of the total is mobilised, meaning that part of the chrome is associated to the crystalline structure of the recalcitrant minerals in the sludge. However, the great difference between the percentage extractable in 0.1 N nitric acid and 10% nitric acid suggests that the most part of this metal is bound to the organic matter.

XRD

Figure 3 shows the spectra obtained after XRD analysis of the original material and the residues obtained after the different treatments. The characteristic reflection of each of the mineral phases identified and the corresponding diffraction angle are shown in Table 6.

Table 6. Phases identified, characteristic reflection and diffraction angle of the XRD spectra obtained for the aerobic sludge

Abbreviation	Mineral phase	Characteristic Reflection (Å)	Diffraction angle (2θ)
K	*kaolinite*	7.15	12.33
C	*Calcite*	3.03	29.50
Q	*Quartz*	3.34	26.70
D	*Dolomite*	2.88	31.07
Fd	*Feldspars*	3.18	27.98
P	*Pyroxene*	2.99	29.78

The mineralogy of the aerobic sludge is basically represented by calcite and quartz, although dolomite, kaolinite and feldspars are also present in substantial amounts. After the treatment with 0.1 N and 10% nitric acid, quartz, calcite and feldespars become predominant. A new phase (pyroxene) was also identified, which, in the other samples, was probably masked by the characteristic peak of calcite. After both acid treatments, the calcite, as expected, was totally dissolved, as was the dolomite, which is more resistant to acid attack. This confirms the complete dissolution of the phases dissolvable in this medium and the release of metals associated to them.

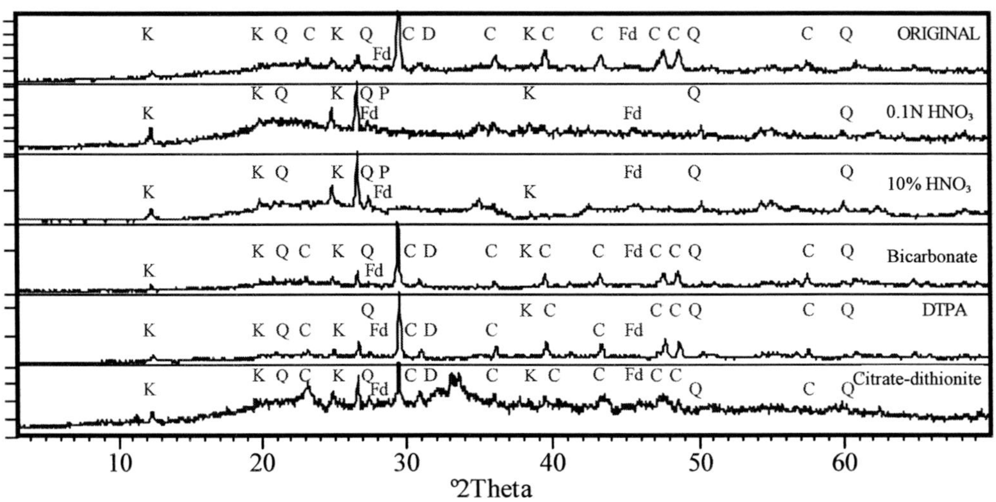

Figure 3. XRD spectra of original aerobic sludge and residues after the different treatments

Treatment with bicarbonate had no effect on the mineralogical composition of the aerobic sludge, which maintained the same composition as the untreated sample. Extraction in bicarbonate medium, a very weak acid, would represent the soluble metal fraction (and probably exchangeable) of the sludges, which is obviously very low in all the sludges studied.

DTPA treatment involved very little change from the original phases, only a slight increase in the proportion of feldspars to the detriment of kaolinite being of note.

No evidence was observed in the original sample of hematites, goetite or other crystalline minerals of iron susceptible to solubilisation in the citrate-dithionite medium. There was therefore very little variation in the XRD diagram after this treatment. The high proportion of metals mobilised in this medium were probably due to their association with non-crystalline iron and manganese oxides, which are difficult to detect by XRD.

Simple Extraction Applied to the Anaerobic Sludge

Figure 4 shows the results obtained after application of the different extractions to the anaerobic sludge.

Of note is the high total content of heavy metals in this sludge, which, in the case of chrome, exceeded the level permitted by Spanish legislation by up to three times. It also had a high total concentration of cadmium, which was very close to the legal limit for application to soils with a pH lower than 7.

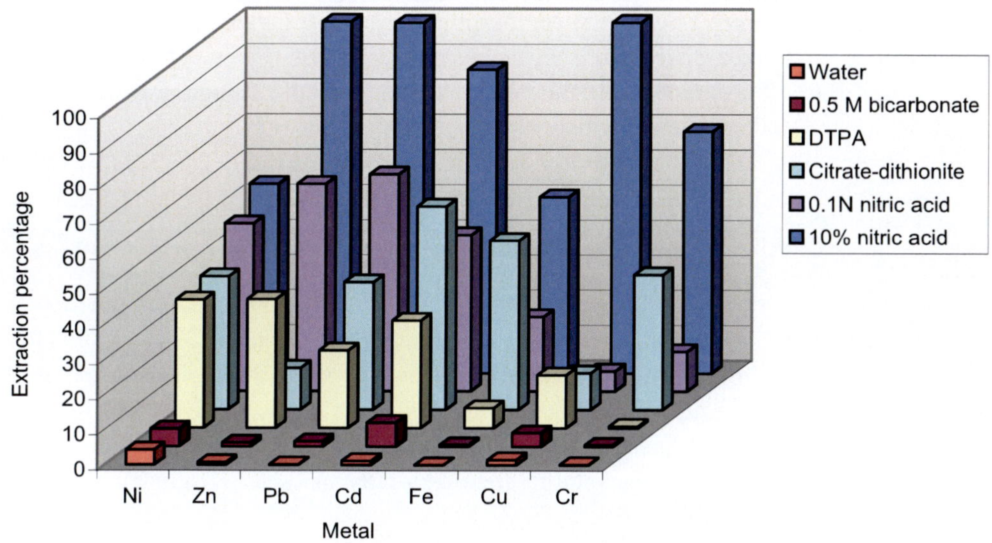

Figure 4. Extraction percentages of each metal after simple extractions applied to the anaerobic sludge

The results obtained for chrome point to its very low degree of mobility in the anaerobic sludge. In the most readily available and bioassimilable fractions, the extraction percentage was practically zero. In reducing conditions the proportion of this metal mobilised was 41% of the total, suggesting that a high proportion was bound to iron and manganese oxides and hydroxides. However, the proportion mobilisable in nitric acid was relatively low, underlining the strong association of this metal to the crystalline structure of the minerals in the sludge. The great difference observed in the amount of chrome mobilised by 10% nitric acid compared with that mobilised by 1N shows that most of the element was associated to the organic matter.

The solubility of cadmium in water and bicarbonate was low, while extraction exceeded 15% in DTPA. This means that about 3 mg/kg of this metal was in bioavailable form, which, given its high toxicity, implies a potential risk of contamination of the soil to which it may be added and a risk that it may enter the food chain. The proportion of cadmium mobilised in reducing conditions was practically zero, and, in strong acid conditions the proportion extracted did not exceed 50%, meaning that part of the cadmium present in this sludge was strongly retained by the mineral fraction most difficult to attack, while another part was bound to the organic matter.

As regards the iron, the low extraction percentage obtained in the most available fractions (water-soluble, extractable in bicarbonate medium and DTPA) is of note. Furthermore, almost 75% of this element is in its free form, not bound to silicates, as can be seen from the high extraction percentage obtained in chelating-reducing conditions. The high percentage of iron mobilised in 0.1 N nitric acid suggests that much of the metal forms sulphides and carbonates. However, in strong acid conditions, less than 75% is mobilised, meaning that a large proportion present in the anaerobic sludge is associated to the silicates it contains.

The mobility of lead in the anaerobic sludge was low because of the low extraction percentage obtained both in the readily available fractions and in reducing conditions. High extraction percentages were only obtained in acidic conditions; indeed, about 82% of the mobilisable lead was released by 10% nitric acid, suggesting that most of this metal is retained in the crystalline structure of the minerals found in the sludge or in the form of sulphides and bound to the organic matter.

As occurred with lead, the mobility of zinc was low in the fractions of greatest availability and even in reducing conditions. Up to 79% was mobilised by 0.1 N nitric attack and almost 100% in strong conditions, which suggests that this metal mainly forms part of the sulphides, carbonates and organic matter.

Nickel was the most available element in the anaerobic sludge, and a high degree of mobility was attained in the most readily available fractions, reaching 23% with DTPA. In reducing conditions, this degree of mobility rose to 48%, suggesting that a large proportion of this metal is associated to iron and manganese oxides and hydroxides. However, with 0.1N nitric acid, only 43% was mobilised, while in strong acid conditions the percentage recovered did not exceed 52%, implying that a large fraction of the nickel is occluded in the most inaccessible part of mineral phase of this sludge.

Alongside nickel, copper was the most soluble element in water and bicarbonate medium, although no more than 5% was extracted. The percentage mobilised in DTPA was 16%, implying a 55 ppm concentration of this element available to plants. However, its mobilisation in reducing conditions was slight. In 0.1 N nitric acid, 58% of the total copper was mobilised and 97% in the more aggressive conditions. This suggests that this element is principally found in the form of sulphides and carbonates or bound to the organic matter of the sludge.

XRD

Table 7 shows the mineral phases identified after XRD analysis of the anaerobic sludge, and the characteristic reflection and diffraction angle corresponding to each phase.

Table 7. Phases identified, characteristic reflection and diffraction angle of the XRD spectra obtained for the anaerobic sludge

Abbreviation	Mineral phase	Characteristic Reflection (Å)	Diffraction angle (2θ)
M	*Moscovite*	10	8.78
C	*Calcite*	3.03	29.50
Q	*Quartz*	3.34	26.70
D	*Dolomite*	2.88	31.07
Fd	*Feldspars*	3.18	27.98
P	*Pyroxene*	2.99	29.78

As can be seen from Figure 5, the mineralogical composition of the anaerobic sludge is mainly made up of calcite (with diffraction angles(2θ) of 29.5 and 39.5°), quartz (with diffraction angles of 26.7 and 20.9°) and dolomite (31.1°); also appearing are moscovite (at 8.98 and 20°) and feldspars (27.95°).

After treatment with 0.1 N and 10% nitric acid, all the calcite and dolomite was dissolved, the peaks corresponding to these phases disappearing from the corresponding diffractogram. The quartz, moscovite and feldspars become predominant, while the peak corresponding to pyroxene at 29.78°, which was masked by calcite in the original sample and in the residues after the other treatments, comes into evidence.

The XRD spectrum of the residue left after the bicarbonate treatment does not differ from that of the original sample, meaning that no crystalline phase of any of the minerals is altered using this medium.

After DTPA and citrate-dithionite treatment, the diffractograms point to an increase in the intensity of quartz, while the others do not change, probably due to alterations in certain organic phases as a result of using these chelating agents.

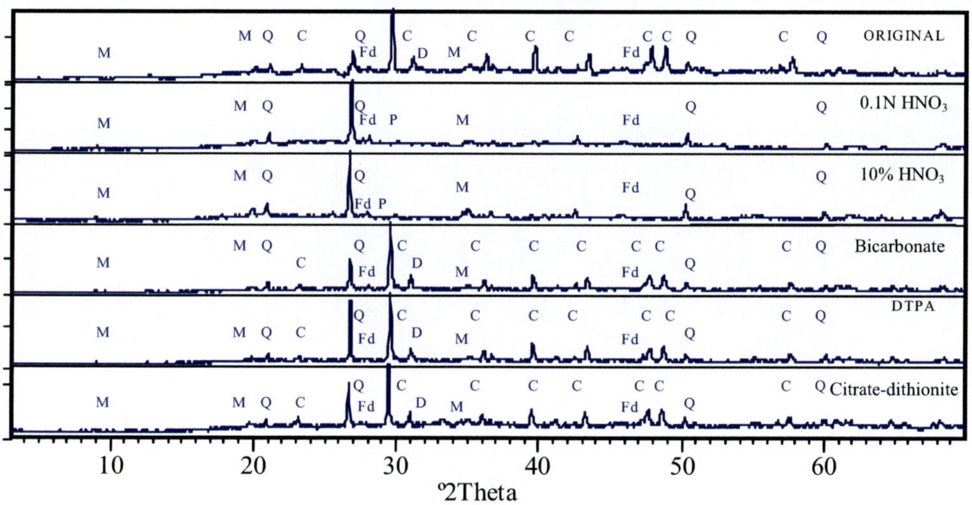

Figure 5. XRD spectra of the original anaerobic sludge and of the residues after the different treatments

Simple Extraction Applied to Non-Stabilised Sludge

Figure 6 shows the results obtained after application of the different simple extractions to the non-stabilised sludge.

The total content of any heavy metal in this sludge did not exceed current legal limits referring to possible agricultural use. In this case, iron followed by zinc and copper were the most concentrated elements.

Nickel showed a high degree of mobility in the most bioavailable fractions, especially in the extraction with DTPA, that exceeded 35%, because part of this element must be in the form of carbonates. The extraction with citrate-dithionite was around 38%, indicating that part of this element was associated to iron and manganese oxides and hydroxides. However, nitric acid attack did not extract more than 55% even in the most aggressive conditions, suggesting that a substantial part was retained in the crystalline structure of the minerals found in the sludge.

After nickel, cadmium was the metal showing the highest degree of bioavailability, about 3% being soluble in water and about 8% in bicarbonate medium, perhaps because part of the cadmium in the sludge was in the form of phosphates. With DTPA, extraction reached 31%, implying a risk of toxicity for any receiving soil. However, the total amount of cadmium in the sludge was low (about 1 mg/kg), meaning that the risk of toxicity was not so great. In reducing conditions cadmium is easily mobilised, suggesting that much of it is retained by the iron and manganese oxides and hydroxides, while the proportion extracted with 0.1 N nitric acid is less than 45%. More than 85% was recovered with 10% nitric acid, so that much must be associated with the organic matter.

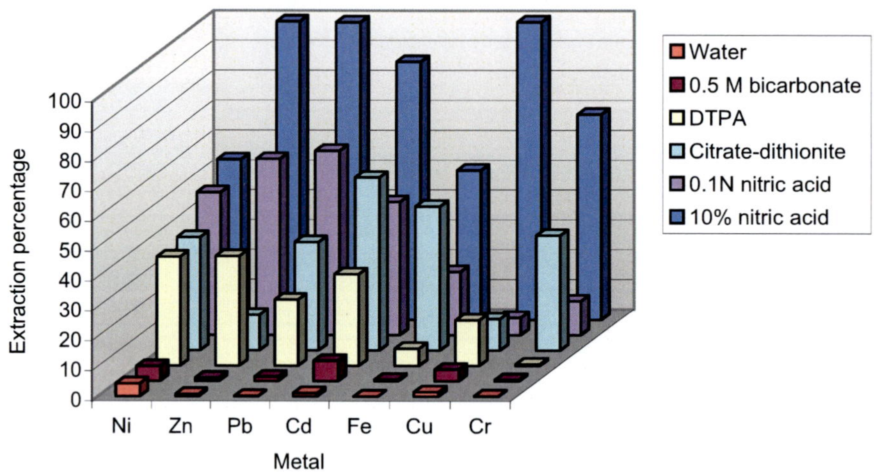

Figure 6. Percentage of each metal extracted with the simple extractions in the non-stabilised sludge

In the case of the non-stabilised sludge, the water solubility of zinc was low, as it was with bicarbonate. However, the bioavailability of this metal exceeded 35%, indicating that 168 ppm were available to plants. Its mobility in reducing conditions was relatively low, about 12% but, despite this, more than 59% was extracted with 0.1 N nitric acid, meaning that much was bound to sulphides. The fact that 100% was recovered with 10% nitric acid indicates that part was associated with the organic matter.

Lead, like zinc, was poorly soluble in water and bicarbonate, and about 22% of the total content is available for plants. The mobility of the lead in reducing conditions was low, while more than 60% was mobilised in 0.1 N nitric acid, which means that a high percentage was in the form of sulphides. As with zinc, close to 100% was recovered with 10% nitric acid, again meaning that much was associated with the organic matter.

Iron showed a low degree of mobility in the three most available fractions (soluble in water and bicarbonate, and extractable with DTPA), not exceeding 6% in any of them. In citrate-dithionite, 48% of the total was mobilised, meaning that most was bound to the silicates. The low extraction percentages obtained in nitric medium confirmed that much of the iron in the non-stabilised sludge formed part of the crystalline structure of the silicates, and not even strong acid conditions using 10% nitric acid was capable of releasing it.

As regards the copper, it showed a low degree of mobility in all the extractions, including 0.1 N nitric acid. However, almost all was the metal was extracted in 10% nitric acid, suggesting it was strongly bound to the OM in the form of stable complexes.

The mobility of chrome in the most bioavailable fractions was practically zero, while in reducing conditions about 40% was extracted, meaning that part, at least, was associated to iron and manganese oxides and hydroxides. In 10% nitric acid, recovery reached 70%, and in 0.1 N only 12%, perhaps because a high proportion was retained by the organic matter, while another fraction was bound to the crystalline structure of the sludge.

XRD

Figure 7 shows the diffractograms of the original non-stabilised material and of the residues after the different treatments, while Table 8 shows the mineral phases identified, their characteristic reflections and corresponding diffraction angles.

Table 8. Phases identified, characteristic reflection and diffraction angle of the XRD spectra obtained in the non-stabilised sludge

Abbreviation	Mineral phase	Characteristic Reflection (Å)	Diffraction angle (2θ)
Mi	*Mica*	10	8.89
C	*Calcite*	3.03	29.50
Q	*Quartz*	3.34	26.70
D	*Dolomite*	2.88	31.07
Fd	*Feldspars*	3.18	27.98

The mineralogy is basically made up of calcite, quartz and dolomite, although the presence of mica and feldspars is also noticeable.

Attack by either of the nitric medium used totally dissolved the calcite and dolomite, their characteristic peaks disappearing from the corresponding diffractograms; as a consequence, quartz predominated, while mica and the feldspars also remained, since these are very recalcitrant materials.

Treatment with bicarbonate had no effect on the original X-rays and the minerals identified showed the same proportions, as can be seen from Figure 7.

Treatment with DTPA, as in the case of citrate-dithionite, led to a substantial increase in the intensity of the quartz, converting it into the principal phase and displacing calcite. However, the spectrum corresponding to the residue left after citrate-dithionite attack showed an increase in background noise, which hindered the complete identification of all the phases present.

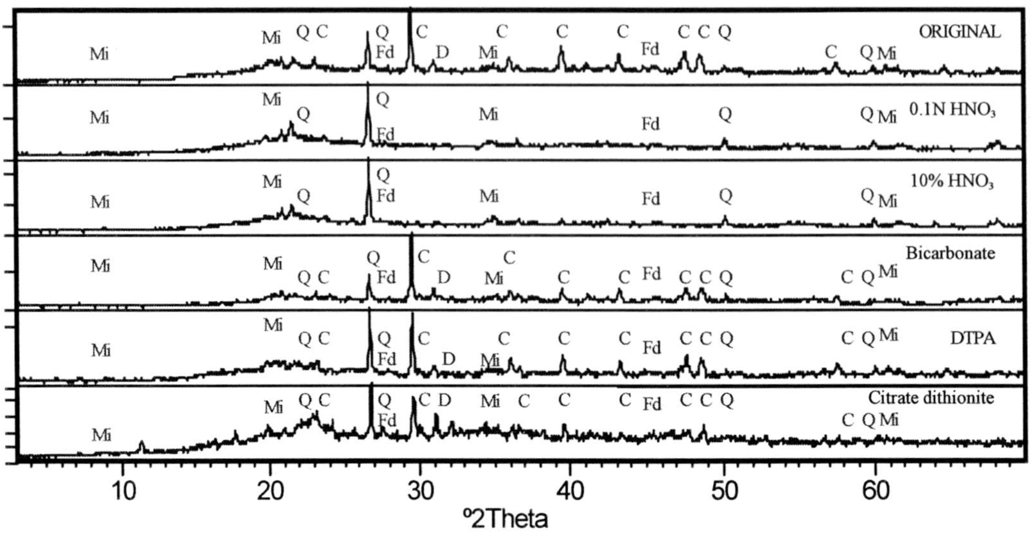

Figure 7. XRD spectra of non-stabilised sludge and residues after the different treatments

Simple Extraction Applied to WSP Sludge

Figure 8 shows the results obtained after the simple extractions were practised in the WSP sludge.

The level of heavy metals in this sludge is lower than the legally permitted values, making it suitable for agricultural use. However, the concentration of cadmium, which exceeds 11 mg/kg, can be considered high, and it must be borne in mind that the forthcoming European Directive will establish a limit of 10 mg/kg in sludges intended for agricultural use.

The availability of cadmium is low since the amount solubilised in water and bicarbonate is close to zero, while DTPA extracts 17% and the reducing medium does not extract more than 5%. However, more than 95% is extracted with 10% nitric acid. The great difference between the mobility of cadmium in dilute nitric acid and that observed after strong acid attack indicates that much of the cadmium is associated with the organic matter in the sludge.

Nickel shows the highest degree of mobility in the WSP sludge, although the low percentage extracted in water (1.7%) and bicarbonate (4%) points to a low availability. The fraction available to plants is around 15%. In reducing conditions up to 60% is mobilised, probably because much is retained by iron and manganese oxides and hydroxides. The percentage extracted with 0.1 N nitric acid is similar to that extracted with 10% nitric acid, probably because the nickel is associated to sulphides and because a substantial amount is in the silicated forms of the minerals found in the sludge (phylosilicates, feldspars and pyroxene).

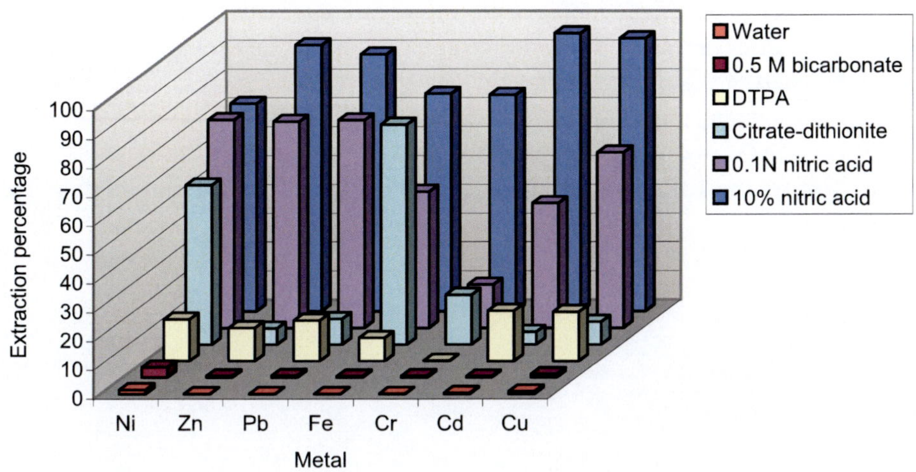

Figure 8. Percentage of each metal extracted after simple extraction applied to the WSP sludge

As regards the zinc content, except in DTPA, the solubility is practically nil in the fractions of greatest savailability. In reducing conditions, for example, no more than 6% is mobilised, while the highest percentage was reached in acid media, 71% being released by 0.1 N nitric acid and 92% by 10% nitric acid, suggesting that most of the metal is associated to the sulphides and organic matter.

Both lead and copper behave in the same way as zinc in this sludge, showing low solubility in water and bicarbonate , while their bioavailability increases to 15% in DTPA. In citrate-dithionite, about 8% of both metals is extracted; in acid conditions, the percentage reaches 72% in the case of lead and 61% in the case of copper, meaning a large part of both is to be found in the form of sulphides and carbonates. Recovery in 10% nitric acid reaches 95% for both elements, so that large proportion must be bound to the orgaic matter and/or associated to the mineral phases, from which they are only released by strong acid conditions.

The availability of chrome is practically nil, this element not being mobilised in any of the most bioavailable fractions. Furthermore, in a reducing medium, the percentage extracted is below 18%, the same as obtained with 0.1 N nitric acid, while 75% is released by 10% nitric acid. This great difference in the amount extracted with the acids indicates that much of the chrome is bound to the organic matter.

Once again iron shows a low degree of bioavailability in this sludge, most being in the form of free oxides and hydroxides and only a small proportion bound to the silicates, as can be gathered from the high percentage of iron mobilised in citrate-dithionite (around 75%). The recovery with 10% nitric acid reaches 75%, while 0.1 N nitric acid mobilises no more than 48%, meaning that the iron is mainly found bound to the organic matter and silicates.

XRD

Figure 9 shows the diffractograms of the original WSP sludge and after the different treatments, while Table 9 shows the mineral phases identified, their characteristic reflections and corresponding diffraction angles. The mineralogy of the WSP sludge is basically made up of calcite and quartz, although the presence of dolomite, illite and feldspars is also noticeable.

Attack by either of the nitric media used totally dissolved the calcite and dolomite, their characteristic peaks disappearing from the corresponding diffractograms; as a consequence, quartz predominated, although illite and the feldspars also remained, since these are very difficult materials to attack.

Table 9. Phases identified, characteristic reflection and diffraction angle of the XRD spectra obtained in the WSP sludge

Abbreviation	Mineral phase	Characteristic Reflection (Å)	Diffraction angle (2θ)
I	*Illite*	10	8.89
C	*Calcite*	3.03	29.50
Q	*Quartz*	3.34	26.70
D	*Dolomite*	2.88	31.07
Fd	*Feldspars*	3.18	27.98

Treatment with bicarbonate had no effect on the X-ray spectra and the minerals maintained their proportions, as can be seen from Figure 9. No differences were seen between the diffractograms obtained from the residues after treatment with DTPA and citrate-dithionite and those obtained for the original WSP sludge.

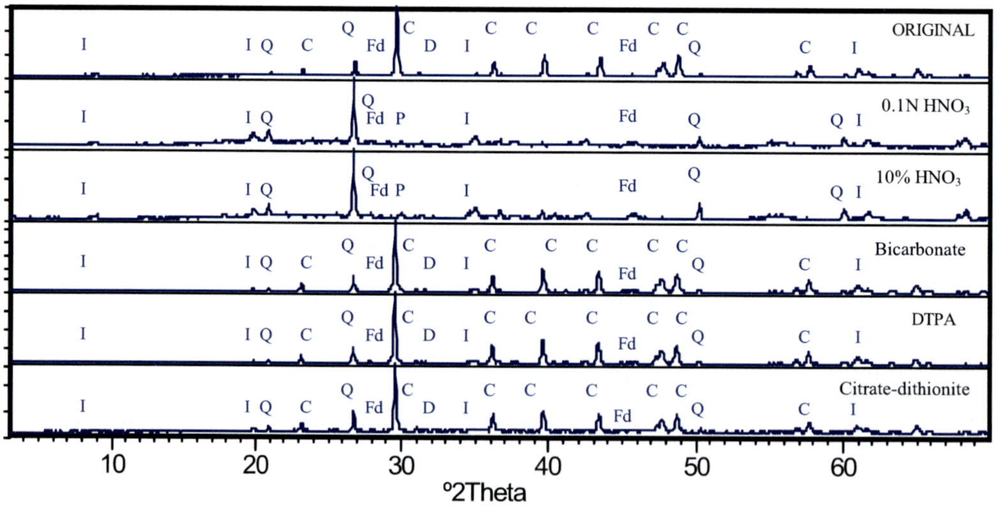

Figure 9. XRD spectra of the original WSP sludge and for the residues after the different treatments

CONCLUSIONS

The heavy metal concentrations measured in the sludges do not exceed the legally established limits, except in the case of chrome in the anaerobic sludge, when the levels almost tripled the permitted values. The other sludges, therefore, seem to be viable natural alternatives for use in agricultural soils since their characteristics are suitable for improving soil fertility.

Extraction with Water

The first thing to note is the low and similar degree of water solubility shown by all the metals in the different sludge types analysed. Nickel is the most available of the metals followed by copper, cadmium and zinc, while lead, iron and chrome mobility is practically nil. The the least mineralised sludges (non-stabilised and aerobic) show the highest degree of extraction.

Extraction with Bicarbonate

The results obtained in this medium are very similar, but slightly higher than those obtained in water. The availability in the rhizosphere of the metals studied is very low, with lead, iron and chrome showing the least mobility in weakly acid conditions.

Extraction with DTPA

In general, the metal fractions show greater bioavailability in the least stabilised sludges, particularly when compared with the anaerobic and WSP sludges. Indeed, the proportion of bioavailable metal varied between 10 and 20% in these last, while more than 30% was seen to be mobilised in the least stabilised samples.

Of note is the low availability of chrome with almost none being mobilised in any of the sludges, including the anaerobic sample, which contained almost three times the legally established values.

Extraction with Citrate-Dithionite

The use of this extraction medium suggested that a substantial fraction of iron is associated to the crystalline structures of the constituent minerals (silicate forms), although much is also present in free forms (oxides and hydroxides).

Chrome and nickel show the highest degree of mobility in the chelating-reducing conditions represented by citrate-dithionite, suggesting that part of these metals is associated to the iron and manganese oxides and hydroxides.

Once again, the metals contained in the most mineralised sludges show the lowest mobility in reducing conditions.

Extraction in Nitric Medium

In general, there is a substantial degree of mobilisation in 0.1 N nitric acid, especially in the case of zinc and lead (in all the sludges studied), meaning that a substantial proportion of these metals is probably in the form of sulphides, carbonates and phosphates.

In strong acid conditions (10% nitric acid), high recoveries were obtained for zinc, lead, copper and cadmium, while only about 70% recovery was attained in the case of iron, chrome and nickel, suggesting that part of these metals is associated to the crystalline structure of the minerals in the sludges.

XRD

The mineralogy of the sludges was predominantly represented by calcite, quartz and dolomite, although with an important contribution of phylosilicates (mica, moscovite and kaolinite) and feldspars.

After acid treatment (with 0.1 N and 10% nitric acid), not only was the calcite totally dissolved, as was to be expected, but also the dolomite, which is more resistant to acid attack, confirming that the attack was complete.

Extraction with bicarbonate did not affect the mineral composition of the sludges, whose components maintained the proportions of the untreated sample.

Treatment with DTPA did not bring about any notable variation with respect to the original phases, and there was only a slight increase in the proportion of feldspars at the expense of the phylosilicates.

Since the original sludges did not show the presence of hematites, goetite or other crystalline minerals of iron susceptible of being solubilised in a citrate-dthionite medium, no great change in the XRD diagrams were observed after this treatment. The high proportions of metals mobilised in this medium are probably due to their association with non-crystalline iron and manganese oxides, which are difficult to detect by XRD.

REFERENCES

Álvarez, E. A.; Mochon, M. C.; Sánchez, J. C. J. and Rodriguez, M. T., 2002. Heavy metal extractable forms in sludge from wastewater treatment plants. *Chemosphere*. 47, 765-775.

Álvarez, E. A.; Mochon, M. C.; Sánchez, J. C. J. and Rodriguez, M. T., 2002. Heavy metal extractable forms in sludge from wastewater treatment plants. *Chemosphere*. 47, 765-775.

Angelidis, M. and Gibbs, R.J. 1991. Heavy metals in sewage sludges: chemical forms and possible availability, in: P.L'Hermite (Ed.), Proceedings of the Treat. Use Sewage Sludge Liq. Agric. *Wastes Symposium, Elsevier*, London, pp. 400-404.

Balmer, P., 2001. Possibilities to improve the quality of wastewater sludges. Proc. IWA Specialised Conference on Sludge. *Entering tne 3rd Millennium*, Tapei, Taiwan, K13-K16.

Berrow, M. L. and Stein, W. M., 1983. Extraction of metals from soils and sewage sludges by refluxing with Aqua Regia. *Analyst.*108; 277-285.

Bish, D. L., 1994. Quantitative X-ray diffraction analysis of soils. En: Quantitative Methods in Soil Mineralogy. Eds. Amonette J.E. y Zelazny, L.W. *SSSA Miscelaneous Publication.* 462 pp.

Bradford, G. R.; Bair, F. L. and Hunsaker, V., 1971. Trace of major elements of soil saturation extracts. *Soil Sci.* 112; 255.

Calvet, R.; Bourgeois, S. and Msaky, J. J., 1990. Some experiments on extraction of heavy metals present in soil. *Intern. J. Anal. Chem.*, Vol. 39, pp 31-45.

Campanella, L.; Cardarelli, E., Ferri, T.; Petronio, B. M. and Pupella, A., 1987. Evaluation of heavy metals speciation in an urban sludge. *The Science of the Total Environment.* 61; 217-228.

Carre, J. and Welte, B., 1986. Speciation des metaux lourds presents dans les boues d'un basin de lagunage naturel. *Envirom. Tech. Letters.*7 ; 351-362

Chlopecka, A. and Adriano, D. C., 1996. Mimicked in situ stabilization of metals in a cropped soil: bioavailability and chemical form of zinc. *Environ. Sci. Technol.* 30 ; 3294-3303.

Davidson, C. M.; Thomas, R. P.; McVey, S. E.; Perala, R.; Litteljohn, D. and Ure, A.M., 1994. Evaluation of a sequential extraction procedure for the speciation of heavy metals in sediments. *Anal. Chim. Acta.* 291; 277-286.

Englande, A. J. and Reimers, R. S., 2001. Biosolids management in sustainable development status and future direction. Proc. IWA Specialised Conference on Sludge. *Entering tne 3rd Millennium,* Tapei, Taiwan, 38-45.

Essington, M. E. and Mattigod, S. V., 1991. Trace element solid-phase associations in sewage sludge and sludge-amended soil. *Soil Sci. Soc. Am. J.* 55; 350-356

Fernández, A.; Pérez-Cid, B.; Fernández, E. and Falqué, E., 2000. Comparison between sequential extraction procedures and single extractions for metal partitioning in sewage sludge samples. *Analyst.* 125; 1353-1357.

Fuentes, A.; Lloréns, M.; Sáez, J.; Soler, A.; Aguilar, M. I.; Ortuño, J. F.; and Meseguer, V. F., 2004a. Simple and Sequential extraction of heavy metals from different sewage sludges. *Chemosphere*, 54; 1039-1047.

Fuentes, A.; Lloréns, M.; Sáez, J.; Soler, A.; Aguilar, M. I.; Ortuño, J. F.; and Meseguer, V. F., 2004b. Phytotoxicity and heavy metals speciation of stabilised sewage sludges. *J. Hazardous Materials,* 108(3), 161-169.

Gambrel, R. P., 1994. Trace and toxic metals in wetlands- A review. *J. Environ. Qual.* 23, 883-891.

García, C.; Hernández, T. And Costa, F., 1991. The influence of composting on the fertilizing value of an aerobic sludge. *Plant Soil,* 136, 269-272.

García-Rizo, C.; Martínez-Sánchez, J. and Pérez-Sirvent, C., 1999. Environmental transfer of zinc in calcareous soils in zones near old mining sites with semi-aridic climate. *Chemosphere*. 39 (2); 209-227.

Hernández, T; Moreno, J. L. and Costa, F., 1991. Influence of sewage sludge application on crop yields an heavy metal availability. *Soil Sci. Nutr.* 37 (2); 201-210.

Hooda, P. S. and Alloway, B. J., 1994. Changes in operational fractions of trace metals in two soils during two-year of reaction time following sewage sludge treatment. *Intern. J. Environ. Anal. Chem.* 57; 289-311.

Lee, D. J.; Spinosa, L. and Liu, J. C., 2002. Towards sustainable sludge management. *Water* 21. December; 22-23.

Legret, M., 1993. Speciation of heavy metals in sewage sludge and sludge-amended soil. *Intern. J. Environ. Anal. Chem.* 51; 161-165.

Levi-Minzi, R.; Sartori, F. and Riffaldi, R., 1981. Caraterization de la fraction organique et de la fraction minerale cristalline des bouse d'epuration. *Proceedings of second European Symposium: Characterization, treatment and use of sewage sludge*. P. L'Hermite y H. Ott (Ed.). Reidel, Holanda. 291-298.

Liang J. and Karamanos, R. E., 1993. DTPA-Extractable Fe, Mn, Cu, and Zn. En " Soil Sampling and Methods of Analysis". Ed. M.R. Carter. *Canadian Society of Soil Science*. Lewis Publishers. 87-90

Lindsay, W. L. and Norvell, W. A., 1969. "Development of a DTPA micronutrient soil test". *Agon. Abstr.*, 84.

Lindsay, W. L. and Norvell, W. A., 1978. "Development of a DTPA test for zinc, iron, manganese and copper". *Soil Sci. Soc. Am. J.* 42, 421-428.

Magoarou, P., 2000. Urban waste water in Europe: What about the sludge? In "Problem around sludge". *Workshop Procedings*, EUR 19657 EN, 9-16, Stresa (I).

Mehra, O. P. and Jackson, M. L., 1960. "Iron oxide removal from soils and clays by a dithionite-citrate system buffered with sodium bicarbonate". Clays and Clay Minerals, Proc. 7th Conf., pp. 317-327. *Natl. Acd. Sci-Natl. Res. Council. Publ.*

Mingot, J. I.; Obrador, A.; Álvarez, J. M. and Rico, M. I., 1995. Acid extraction and sequential fractionation of heavy metals in water treatment sludges. *Environ. Technol.* 16, 9; 869-876.

Moreno, J. I. S., 1985. Lodos de depuradora urbanas: Caracterización y aplicación agrícola. *Tesis Doctoral*. Universidad de Murcia.

Moreno, J. L., 1997. Uso de composts de lodo de depuradora para la mejora de la calidad de los suelos de zonas áridas. Efecto de su contaminación metálica. *Tesis Doctoral*. Universidad de Murcia.

Olsen, S. R.; Cole, C. V.; Watanabe, F. S. and Dean, L. A., 1954. "Estimation of available phosphorus in soils by extraction with sodium bicarbonate". *U.S. Dept. Agr. Circ.*, 939.

Olsen, S. R. and Dean, L. A., 1965. " Methods of soil Analysis", Parte 2. C.A. Black(Ed). *Amer. Soc. Agronomy, Inc.*, Madison, Wis. 1044-1045.

Olsen, S. R. and Sommers, L. E., 1982 Phosphorus, En Methods of soil analysis. Part 2. 2nd Ed. *American Society of Agronomy*, Madison, WI.

Ortuño, J. F., 1993. Intercambio de fósforo entre agua y sedimentos generados en una laguna profunda de estabilización. *Tesis Doctoral*. Universidad de Murcia. España.

Pérez-Cid, B., Lavilla, I. and Bendicho, C., 1999. Comparison between convencional and ultrasound accelerated Tessier sequential extraction schemes for metal fractionation in sewage sludge. *Fresenius J. Anal. Chem.* 363; 667-672.

Petruzzelli, G., 1989. Recycling wastes in agriculture: heavy metal bioavailability. *Agric. Ecosyst. Environ.* 27; 493-503

Pueyo, M; Rauret, G.; Luck, D.; Yli-Halla, M.; Muntau, H.; Quevauville, Ph. and López-Sánchez, J. F., 2001: Certification of the extractable contents of Cd, Cr, Cu, Ni, Pb and Zn in a freshwater sediment following a collaboratively tested and optimised three-step sequential extraction procedure. *Journal Environmental Monitoring*, 3-2; 243-250

Ross, G. J. and Wang, C., 1993. Extractable Al, Fe, Mn, and Si. En " Soil Sampling and Methods of Analysis". Ed. M.R. Carter. *Canadian Society of Soil Science*. Lewis Publishers. 239-246.

Scancar, J.; Milacic, R.; Strazar, M. and Burica, O., 2000. Total metal concentrations and partitioning of Cd, Cr, Cu, Fe, Ni and Zn in sewage sludge. *Science of Total Environment.* 250 (1-3); 9-19.

Schoenau, J. J. and Karamanos, R. E., 1993. Sodium Bicarbonate-Extractable P, K, and N. En " Soil Sampling and Methods of Analysis". Ed. M.R. Carter. *Canadian Society of Soil Science*. Lewis Publishers. 51-58.

Sirviera, D. J. and Sommers, L. E., 1977. Extractability of copper, zinc, cadmium and lead in soils incubates with sewage sludge. *J. Environ. Qual.*, 6; 47-52.

Smith, S.R., 1996. Agricultural Recycling of Sewage Sludge and the Environment. *CAB International*, UK.

Soon, Y. K. and Abboud, S., 1993. Cadmium, Chromium, Lead, and Nickel. En " Soil Sampling and Methods of Analysis". Ed. M.R. Carter. *Canadian Society of Soil Science*. Lewis Publishers. 101-108.

Spinosa, L., 2001. Evolution of sewage sludge regulations in Europe. *Water Sci. Technol.*, 44(10), 1-8.

Su, D.C. and Wong, J.W.C., 2003. Chemical speciation and phytoavailability of Zn, Cu, Ni and Cd in soil attended with fly ash-stabilized sewage sludge. *Environ. Int.* 1060, 1-6.

Vidal Otón, J., 2002. Evaluación de los principales procesos de degradación en fluvisoles calcáricos de la Huerta de Murcia.*Tesis Doctoral*. Universidad de Murcia.

Welte, B. and Montiel, A., 1985. Comparaison de différentes méthodes d'attaque des sédiments et boues de station d'épuration. *Trib. Cebedaeau.* 503 (38); 3-13.

Wong, J. W. C.; Li, K.; Fang, M. and Su, D. C., 2001. Toxicity evaluation of sewage sludges in Hong Kong. *Environ. International*, 27; 373-380.

Zufiaurre, R.; Olivar, A.; Chamorro, P.; Nerín, C. and Callizo, A., 1998. Speciation of metals in sewage sludge for agricultural uses. *Analyst.* 123; 255-259.

In: Water Pollution: New Research
Editor: A. R. Burk, pp. 29-53

ISBN: 1-59454-393-3

Chapter 2

BROWN COLORED WASTEWATERS FROM MOLASSES FERMENTATION PROCESSES

Mónica Coca, Mar Peña [1] and Gerardo González
Departamento de Ingeniería Química y Tecnología del Medio Ambiente
Universidad de Valladolid
Valladolid, Spain

ABSTRACT

Environmental problems in beet molasses fermentation factories are mainly related with the production of large quantities of a high strength effluent known as vinasse. This paper reports on the main characteristics of both beet molasses and vinasse and summarizes methods for using and treating molasses processing effluents. Suitable solutions have to be proposed due to the more restrictive legislative rules. Anaerobic digestion is an economically viable and widely used vinasse treatment scheme. Anaerobic digestion is usually followed by an aerobic treatment to meet standard discharge values. However, these treatments are not able to degrade colored compounds in the effluents. Therefore, the anaerobic-aerobically treated effluent has negative environmental impact if released into surface and ground waters due to its brown color. Much of this paper focused on characterize the colorants present in the molasses processing streams and on examining those treatments which are expected to degrade the non-biodegradable colored compounds. Among them, chemical oxidation with ozone is a promising technique for removing color from both the vinasse and the anaerobic-aerobically treated effluent without producing by-products that need further treatments.

---

[1] Departamento de Ingeniería Química y Tecnología del Medio Ambiente, Universidad de Valladolid
Pº Prado de la Magdalena s/n, 47011 Valladolid, Spain
E-Mail address: pena@iq.uva.es

EFFLUENTS FROM MOLASSES FERMENTATION PROCESSES

Molasses, the main by-product of sugar beet processing, is widely used as cheap raw material in fermentation processes, such as alcohol production, baker's yeast fermentation and amino acid production. The effluents from molasses-based industries, which are known as vinasse, present high organic load and dark brown color. Vinasse is usually subjected to an anaerobic digestion for removing organic matter and producing biogas, which can be used as fuel for the molasses fermentation process. After the anaerobic digestion, an aerobic treatment is added to meet discharge norms. However, the colored compounds cannot be degraded and remain in the aerobic effluent. The discharge of colored molasses wastewaters into waterways is detrimental to aquatic quality due to the reduction in the photosynthetic activity, leading to lower oxygen availability. Environmental problems are one of the most important concerning society, especially the wastes produced by industrial activities. Molasses wastewater effluents are complex systems containing, in addition to coloring polymers formed during the sugar production process, a high concentration in inorganic salts. The complexity of the wastewater matrix makes difficult the development of effective treatments schemes for removing the recalcitrant colorants and for preventing the environmental impact of colored molasses effluents.

The first part of this chapter aims to present an overview of the main characteristics of both molasses and vinasse, including information about the structures and properties of colorants that may be responsible for the color. The compilation is essential to enhance understanding of the nature of coloring matter and to develop effective treatment schemes for removing color from effluents. After a literature survey concerning molasses wastewater characteristics, a compilation on the processes developed to treat molasses effluents, follows. The second part of the chapter reports on the characteristics of the effluents from a beet molasses alcoholic fermentation factory, including the characterization of the colorants present in the wastewaters. There are no studies attempting to characterize the coloring matter. Gel permeation chromatography (GPC) is a suitable analytical technique for determining both the molecular weights of colored compounds in molasses effluents and their UV/Vis spectra by using a photodiode array (PDA) as detector. By comparing the spectra of colorants in effluents with those of synthetic colorants previously prepared can be performed the characterization of coloring impurities. Finally, a study analyzing the efficiencies for removing color of a chemical oxidation process with ozone was performed. During recent years ozonation processes have been successfully applied to degrade colored organic compounds. However, little research in the ozonation of highly polluted molasses wastewaters has been reported. The main parameters influencing the efficiencies reached during the ozonation of beet molasses wastewater were thorough studied due to the advantages of this treatment compared with other decolorization processes such as physicochemical treatments and biological treatments using selected microorganisms.

BEET MOLASSES

The main by-products of sugar beet manufacture are pulp and molasses, but molasses is the most valuable. From a technological viewpoint, molasses is the run-off syrup from the final stage of crystallization, from which no additional sugar can be obtained economically by further crystallization. Beet molasses contains sugars (mono- and oligosaccharides) and those nonsugars compounds from sugarbeet that remain in the juice after purification, together with compounds formed during the sugar beet processing. The exact composition of molasses depends on the original agricultural practices, the region in which the beets are grown, the weather during the growing period, the harvesting method and the technology used at the sugar beet factory [1]. Approximately 80% of color of molasses is formed in the sugar house, when the operation conditions favor the formation of colorants.

Beet Molasses Characterization

The dry matter content in beet molasses is about 74 to 77% and the level of total sugars ranges from 47 to 50%. About 30% of the total mass is composed of nonsugar organic matter. The nonsugar substances in molasses are divided into nitrogen-containing organic nonsugars, about 40% the total nonsugar mass, inorganic, about 33%, and nitrogen-free organic nonsugars, about 20% [1, 2].

Beet molasses contains mainly sucrose (48%) but also invert sugars, e.g., fructose (0.6%) and glucose (0.4%). Small amounts of kestoses as fructosylsucroses, raffinose and galactinol (α-galactosyl-m-inositol) are also present (about 1%). Because virtually all the mono- and oligosaccharides in beet molasses are fermentable by yeast and other microorganisms, beet molasses is preferably used for fermentation processes.

The main nitrogen-containing organic compound is betaine (3 to 4% of beet molasses). More free and bound amino acids are found, such as glutamic acid, aspartic acid and pyrrolidone carboxylic acid (total of 2 to 3% by mass of beet molasses). The pyrrolidone carboxylic acid content results from the saponification of the glutamine during juice purification [3]. About 30% of the total nitrogen content in molasses is formed by nitrogen containing colored compounds not clearly identify.

Beet molasses contain large amounts of mineral matter. Potassium, the major constituent cation (3.6%), is an essential nutrient for some microorganisms and animals. Potassium comes from the sugarbeet plant and it is not removed during the sugar manufacture since potassium salts are soluble and accumulate in the final molasses, accounting for up to 70% of the total ash content. The next most common cation of the ash in beet molasses is sodium (0.9%), followed by calcium (0.4%) and magnesium (0.05%). If decalcification of thin juice is applied, much lower calcium levels can be found. The sodium ion content is dependent on the sodium content of the soil and the decalcification process, using either ion exchangers or soda ash to maintain basic pH values. The main inorganic anionic components are chloride (1.2%), sulfate (1%) and nitrate (0.25%). Traces of phosphate together with rather low contents of nitrite are also found. The presence of nitrite is undesirable for yeast producers, because it interferes with the growth of yeasts.

Beet molasses may contain up to 4% of organic acids, predominantly lactic acid (up to 1.7%), followed by maltic, citric, tartaric, succinic and oxalic acids. Lactic acid is formed by alkaline invert sugar destruction during juice purification and/or via microbiological activity during juice extraction. Although sugar beet is relatively low in vitamins, compared with other plants, small quantities of vitamins can pass through the process, ending up molasses, mainly inositol, nicotinic acid, choline, biotin, and pyridoxine (B6). Vitamins can provide benefits in fermentation and animal feed.

Beet Molasses Utilization

Due to the high content in fermentable carbohydrates, molasses has been widely used since the 17th century in rum and ethanol production. Since the 20th century, molasses has also important industrial applications in the manufacture of bakers' yeast, citric acid and amino acids. Major products from molasses processing are ethanol, citric acid, L-lysine, L-lactic acid, L-glutamic and β-lactam antibiotics. Small volume products include itaconic acid, L-tryptophan, dextran and polyhydroxybutyrate [4]. Bioethanol is manufactured on a large scale for use as liquid transportation fuel. L-(+)-lactic acid is developing a role as food additive as well as key chiral chemical in the synthesis of agrochemicals and pharmaceuticals. Fermentation processes may require the addition of mineral acids to achieve a certain pH value, e.g., pH about 4.5 for optimum growth of yeast. Because of the buffering capacity molasses may require ten times as much acid to reach a certain pH value. Acidification with sulfuric acid produces the precipitation of calcium as insoluble sulfates. Otherwise these would interfere with fermentations especially in lysine and L-glutamic acid production, as well as yeast manufacture. Acidification also promotes the inversion of sucrose to invert sugars, which are more quickly fermented.

Because of their sugar and protein content, beet molasses are also suitable for livestock feed improving texture, palatability and feeding values. Molasses can also be a source of betaine, nutrient for fish feeding. Due to the high viscosity and adhesive properties, molasses can be used for physical/chemical binding of particles and agglomerates. The binding molasses property has been used for many years in the manufacture of animal feed blocks and in cementing coal particles together into briquettes for industrial and domestic use. However, the high nitrogen content of beet molasses produces ammonium salts, which may yield white deposits on combustion. Moreover, the high levels of potassium cause the burning to be too rapid. For these considerations the utilization of beet molasses for coal briquetting has been excluded.

In some economic circumstances, it can be advantageous to extract more of the sugar from molasses in crystalline form using calcium, osmosis, ion exchange, ion exclusion and electrodialysis [5]. The nonsugars components from these processes are concentrated to 60-75% dry matter content and sold as residual molasses non suitable for fermentation due to the low sugar content. Rather than disposing of them by incineration or with the wastewater, residual molasses from desugarizing processes can be a source of livestock feed, betaine and amino acids. Moreover, in the course of sugar desugarizing using chromatographic separation, it is possible to obtain betaine, inositol, amino acid mixtures and individual amino acids, improving the economy of sucrose recovery.

VINASSE FROM BEET MOLASSES FERMENTATION PROCESSES

As mentioned above, most molasses (about 90%) produced by the sugar beet factories is used as raw material in fermentation processes. Fermentation factories using beet molasses produce very large amounts of dark brown colored wastewater that is known as vinasse. For instance, about 10 to 15 liters of vinasse are produced for every liter of ethanol in distilleries [6]. Vinasse is difficult to dispose of, involving an important problem concerning the economic viability of molasses fermentation factories. A great effort has been invested during the closing 25 years of the last century to transform the vinasse into interesting, valuable and marketable products.

Beet Vinasse Characterization

The production and characteristics of vinasse are highly variable and dependent on the molasses fermentation process. Most nonfermentable organic compounds in molasses end up in the fermentation effluents. The range of dry matter is between 58-65%. The chemical oxygen demand (COD) value is considerably large, about 60-100 g/L. The high biological oxygen demand (BOD) values, about 30-60 g/L, indicate that organic matter is mainly formed by biodegradable compounds. There is a relative high protein equivalent, 18-22%, derived from the original nitrogen in beet molasses. Vinasse is especially rich in betaine, about 15-20 g/L, which is not assimilated by yeast. Colored compounds are not assimilated by yeast either, remaining in the effluent. The mineral matter content is about 33-38%, due to the presence of ions, mainly potassium, sulfate and chloride. The potassium content as K_2O is considerably high, about 5-10 g/L. The high organic matter content of vinasse means that before disposing of the effluent it is necessary to go through treatment stages and to study new technologies to utilize vinasse.

Beet Vinasse Utilization

Fertilizer

Vinasse is a potential fertilizer due to the high nitrogen and potassium levels. The nitrogen of vinasse is almost completely in organic form, such as amino acids, glutamic acid and betaine. Beet vinasse is widely used as natural fertilizer for sugar beet crops, increasing both yield and soil fertility without reducing quality [6]. Concentrated vinasse can also be used in organic farming. However, the use of concentrated vinasse as fertilizer may have some limitations due to its high salinity, high density, about 1.3 Kg/L, and low phosphorus content. These problems may be overcome by co-composting the vinasse with other solid agricultural wastes, such as cotton gin wastes, grape wastes and olive wastes [7, 8], using windrows and static pile systems. Composts obtained from vinasse have been reported to have positive effects on the nutritional status, growth and yield of sugar beet, cotton and corn. Vinasse may also be mixed with the organic fraction of the urban solid wastes to produce high quality compost [9]. The mixture in a ratio of 1:3 is subjected to a fermentation-controlled process for the production of compost. Due to the nutrient content of the vinasse, the fermentation can be completed within 30-50 days in pile or tunnel plants in which the temperature increases to about 70 °C.

The concentration of vinasse for being used as fertilizer takes place in multiple evaporators. The final solid content is limited by the risks of crystallization of large quantities of potassium as potassium sulfate. The requirements in energy are significant and they can negatively impact the energy balance of fermentation factories. To limit the formation of potassium sulfate crystals during the concentration stage or during the storage of concentrated vinasse, it is recommended to acidify the fermentation must with hydrochloric acid instead of sulfuric acid. Membrane separation has also been employed for concentration [10]. While energy consumption is less than for evaporation, membrane fouling is problematic and low molecular weight organics pass through the membranes, reducing the potential recycling of permeate in the fermentation process.

Extraction of Potassium Salts

The excessive potassium content of concentrated vinasse is the main limiting factor for the utilization of vinasse in animal feed due to its laxative effect. Therefore, potassium has to be previously removed if the final use of vinasse is as feeding stuff. Sulfate ions usually need to be added so that the precipitation can take place. The addition of sulfate as ammonium sulfate increases the apparent protein level of concentrated vinasse, rising up to 50% in some products. Some processes for precipitating the potassium as potassium sulfate have been proposed [6]. Concentration-crystallization processes require the concentration of vinasse under vacuum to about 78% solids. Calcium sulfate crystals are firstly precipitated. The use of evaporator constituted of 4 forced circulation crystallizers favors the production of good quality crystals with a very high yield. The crystallized salts can be separated by ordinary filtration. The use of an elutriating leg can also be possible, due to the higher density of crystals compared with organic matter. However, potassium extraction is not a feasible process since feed market is not economically valuable.

Anaerobic Digestion

Anaerobic digestion has been considered as the most effective and economic alternative for using beet vinasse due to its high organic matter content. Anaerobic digestion is able to reduce the organic load and convert a significant portion of the organic matter (>50% as COD) to biogas [10]. Methane production depends on the organic load and characteristics of vinasse being about 0.35 m^3 CH_4/Kg degraded COD [6]. The biogas, with methane content of about 60-65%, can be burnt to produce steam or electricity. Methane production can cover almost half of the energy requirements of fermentation factories. Sludge formation throughout anaerobic digestion is minimal and nutrient demand is low, favoring economic viability. In spite of its effectiveness for reducing the organic matter content, the effluent coming out the anaerobic digestion is still loaded in organic compounds and nonbiodegradable colorants.

Colorants Formed During Sugar Beet Processing

Color formation during sugar manufacture is an important factor affecting the quality of both sugar and molasses. Color in molasses causes diminution of the assimilable N and formation of nonfermentable compounds. Several compounds take part in the highly complex reaction pathways resulting in the formation of high molecular weight products of intense color. Two main groups of colorants can be considered: colored compounds from the beet root, mainly melanins, and those formed from sucrose during reactions taking place at practically all stages of sugar manufacturing, which can be classified in melanoidins, alkaline degradation products of hexoses (HADPs) and caramels.

Melanins

The enzymatic colored compounds from the beet root are known as melanins. These compounds are related with catalytic reactions of phenolic compounds, such as tyrosine and dihydroxilphenilalanine. Enzymes involved are polyphenol oxidases, containing copper in the active center and catalyzing electron transfer from phenols to oxygen, forming a diphenol and water as products. Tyrosinase is the enzyme in beet juices that catalyzes the oxidation of tyrosine to dopa and subsequently to an orthochinone. The oxidation and polymerization of subsequent products lead to the formation of melanins. Melanins formation takes place mainly during the extraction stage. Melanins are normally insoluble and thus, can be removed during the juice purification. Therefore, such as colorant is not considered a great technological problem.

Melanoidins

Melanoidins are formed as a result of polymerization reactions of highly reactive intermediates formed during the Maillard reaction. The Maillard reaction is initiated by a condensation reaction between reducing saccharides and amines or amino acids. The Maillard reaction is not a single reaction pathway but a multitude of consecutive and parallel chemical reactions resulting in a wide range of reaction products. Some may be formed in considerable extent but many only in trace amounts, which nevertheless can play an important role in the characteristics of the product mixture, specially the color. The main monosaccharides in sugar beet processing are glucose and fructose, whereas the most important amino compounds are lysine, γ-aminobutyric acid and glutamine that is present in beet juices at high concentration. The composition of melanoidins depends on the reaction conditions, mainly temperature, heating time, pH value, water content and the nature and concentration of reactants (11-14). It has been assumed that under conditions of high dry substance content, elevated temperature and residence time, Maillard reaction proceeds much faster. Therefore, the Maillard reaction is of major importance in evaporation and crystallization stages. With increasing reaction time, the molecular weight and the unsaturation increase. The color intensity also increases with the polymerization degree.

Alkaline Degradation Products of Hexoses

Monosaccharides in aqueous alkaline solutions undergo both reversible and irreversible transformations resulting in the formation of enediol anions, which are considered as intermediates in the isomerization reactions of monosaccharides as well as intermediates in alkaline degradation reactions. Several subsequent reactions contribute to the broad spectrum of end products. The products of alkaline degradation reactions can be divided in two groups according to the number of carbon atoms [15]:

$\leq C_6$ carboxylic acids, colorless compounds, mainly lactic acid but saccharinic, formic, acetic and oxalic acids are also formed and pass through the process to molasses.

$>C_6$ carboxylic acids which nature and structure have not been elucidated.

The products other than $\leq C_6$ can polymerize to form high molecular weight colorants. The rate and extent of alkaline degradation reactions are favored by high hydroxyl ion concentrations, elevated temperatures and concentrated monosaccharide solutions. The presence of divalents cations, such as calcium and magnesium, also accelerates the degradation of monosaccharides. For these reasons, the formation of degradation products takes place mainly during the juice purification stage.

Caramels

Caramelization reactions are degradation reactions of sugar, caused by heating solid sugar and concentrated sucrose syrups at temperatures above 210°C. As the reactions described earlier, very complex processes take place, resulting in the formation of high molecular weight colorants, which chemical structure is not satisfactorily known. Sugars should first undergo intramolecular rearrangements [16]. Depending on the time and temperature, yellow or brown solutions are obtained. α-Heterocyclic, alicyclic or intramolecular H-bonded ring compounds are regarded as important structural elements. In addition, the colored compounds are supposed to exhibit elements of pyran structure. Caramelization occurs in beet sugar production when sugar surfaces are heated strongly, e.g., evaporation stage. However, caramel formation from sucrose is assumed to play a minor role in processing due to the lower reaction rate in comparison with melanodin formation.

Treatment of Wastewater from Molasses Fermentation Processes

Treatments for Removing Organic Matter

As previously mentioned, the most widely accepted and economically viable scheme for treating vinasse involves anaerobic digestion, because it is able to reduce organic matter and produce biogas. In addition, the production of sludge and the requirement of nutrients are considerably lower than those of the aerobic treatments. The anaerobic digestion includes different stages for transforming organic matter into carbon dioxide and methane: (a) a hydrolysis phase of high molecular weight compounds, such as polysaccharides, proteins and lipids, by hydrolytic bacteria, (b) conversion of simpler substrates, such as monosaccharides and amino acids, into volatile fatty acids (VFA) by acidogenic bacteria, (c) conversion of fatty acids into acetate by acetogenic bacteria and (d) a final stage of methane production from acetate, H_2 and CO_2 by methanogenic bacteria. As vinasse is rich in fermentable components, the acidogenic phase is very active and it leads to a high volatile fatty acid concentration. The accumulation of fatty acids produces a reduction in the pH, which would inhibit methanogenesis. For this reason, the acidogenic and methanogenic stages are carried out in different digesters. Thermophilic anaerobic digestors achieve similar efficiencies in removing organic matter and producing methane compared with mesophilic treatments, but at almost twice of organic loading rate. The application of thermophilic anaerobic digestion may improve process economics, since smaller reactors and less cooling are required.

If the molasses fermentation process is carried out adjusting the pH with sulfuric acid, the high sulfate content of vinasse may produce some problems in the biodigestion. The sulfate reducing bacteria form more toxic sulfide at the expense of methane production and it leaves the reactor as either sulfide in the liquid effluent or hydrogen sulfide in the biogas. Sulfide inhibits anaerobic bacteria at concentrations of 200 mg S^{-2}/L [6] and it also contributes to odors and corrosiveness. High reactor sulfide levels can be mitigated maintaining an appropriate COD/SO_4^{-2} ratio in the reactor or adding soluble Fe^{+3}, which promotes precipitation of ferrous sulfide [10]. Upflow anaerobic sludge blankets (UASB) are usually used as reactors. The liquid circulates from the bottom to the top and the biomass is mobile. Due to a phenomenon of flocculation and agglomeration, the biomass is in the form of granules in fluid state.

The effluent from the anaerobic digestion usually requires further treatments to meet discharge norms. The anaerobic digestion may be followed by an aerobic treatment that requires energy for transferring oxygen and produces large quantities of sludge requiring further disposal.

Treatments for Removing Color

Colorants in molasses effluents are hardly degraded by anaerobic-aerobic biological treatments and the colored compounds remain in the aerobic effluent, being necessary other treatments for removing colorants before being disposed of. The discharge of highly colored effluents into waterways can reduce the sunlight penetration, which produces the reduction of oxygen levels necessary for the photosynthetic activity. The death and decay of aquatic plants increases the oxygen demand and causes eutrophication.

The methods most widely studied for removing color from molasses fermentation effluents include flocculation-coagulation processes, biological treatments using selected microorganisms and chemical oxidation processes. Other processes such as supercritical water oxidation [17] and membrane processes [18] have also been studied. Supercritical oxidation of vinasse mixed with municipal excess sludge was carried at elevated temperatures of 673-773 K, resulting in rapid total organic carbon (TOC) reduction in few minutes reaction. When microfiltration in series with nanofiltration was applied to decolorize anaerobically digested beet molasses effluents, yields on color and COD removal of about 89 and 72%, respectively, were achieved. Pretreatment of the effluent, such as coagulation and coarse filtration, was required. None of these processes have been attempted at large scale and they cannot be considered as economically viable alternatives for vinasse treatment.

Coagulation-Flocculation Processes

Different inorganic salts, such as aluminum salts, iron salts and calcium salts, inorganic flocculants based on aluminum polychloride and cationic polymers were used for decolorizing both beet vinasse and anaerobic-aerobically treated vinasse [19]. Aluminum salts were able to remove 86% of color from the biologically treated effluent after adjusting the pH at 5. The efficiencies in color removal were similar when adding either the inorganic flocculant or the cationic polymer, but at the original pH value of the wastewater, about 8.5. The percentage of COD removed was of about 55%. Regarding raw vinasse, color removal was lower than 3% for all the reagents tested. In another study, a polymer of ferric-hydroxy-sulfate was used in the flocculation and coagulation of both vinasse and anaerobically digested vinasse from cane molasses processing resulting in a 37 and 87% color reduction, respectively [20]. Physicochemical processes have disadvantages due to the high consumption of chemical reagents and the high volume of solid sludge produced, requiring further management.

Biological Color Removal

During recent years biological color removal has received considerable attention. Many researches have tried to isolate microorganisms able to decolorize molasses processing effluents. The bacterium *Pseudomonas flourescens*, isolated from a soil sample contaminated with molasses, was able to decolorized diluted molasses wastewater samples up to 76% in four days at 30°C under nonsterile aerobic conditions [21]. Immobilized cells could be reused but decolorization activity was found to decrease in subsequent cycles. A strain MD-32 isolated from a soil sample, belonging to the genus *Bacillus* and closely resemble *B. smithii*, was able to decolorize 35% of molasses pigment within 20 days at 55°C under anaerobic conditions, but it was not able to produce decolorization under aerobic conditions [22]. A strain No. BP103 of acetogenic bacteria, selected among 170 strains of acetogenic bacteria, was cultivated in anaerobically digested molasses wastewater, showing a decolorization yield of about 76% when cultivated at 30°C for 5 days after supplementing glucose, yeast extract and nutrients [23]. Without nutrients supplement, the decolorization yield decreased up to 44%. A yeast strain isolated from fruits and identified as *Citeromyces* sp. WR-43-6, showed a decolorization yield of 69% when cultivated at 30°C for 8 days in a molasses pigment solution after supplementing glucose and nutrients [24].

Rhizoctonia sp. D-90 was applied for treating vinasse in a fed-batch system but the maintenance of the mycelium for a long period was difficult due to bacteria contamination [25]. To study the aerobic degradation of vinasse, diluted to 50%, from an alcoholic fermentation factory the following fungi were used: *Penicillium sp.*, *Penicillium decumbens, Penicillium lignorum* and *Aspergillus niger*. [26]. *P. decumbens* showed the maximum decolorization efficiency with a percentage of color removed of 40% after the fourth day of treatment. *P. decumbens* was also able to reduce phenolics, which substantially improved anaerobic digestion. Anaerobic-aerobically treated beet vinasse was decolorized under aerobic conditions by the white rot fungus *Trametes versicolor* [27]. The color removal efficiency was of about 82% after three days of culture. The addition of sucrose as carbon source and nutrients is required. The white rot fungi *Trametes versicolor* was also applied to decolorize anaerobically digested cane vinasse, achieving 71.5% color removal after the addition of glucose [28]. The white-rot fungus *Phanerochaete chrysosporium* was able to decolorize solutions of synthetic colorants, such as melanoidin, caramel and HADP's, with yields of 74, 87 and 80%, respectively [29]. The white rot fungus *Coriolus hirsutus* exhibited a strong ability to decolorize melanoidin-containing wastewaters exempt from organic nitrogen [30]. Therefore, a previous activated sludge pretreatment of the wastewater is required to convert organic nitrogen into inorganic nitrogen.

As can be seen from the literature, when microorganisms were applied to decolorize anaerobically digested vinasse, the percentage of color removal was higher than that obtained with raw vinasse. Most of these microbial decolorization processes required effluent dilution and the addition of nutrients and glucose as carbon source for optimal activity. It is also difficult to maintain the color removal activity in subsequent cycles. These processes has not been applied at full-scale and cannot yet be considered a well-established technology.

Chemical Oxidation Processes

During the last years chemical oxidation processes using ozone, hydrogen peroxide and Fenton´s reagent, in single or combined action, have been successfully applied to the treatment of specific compounds such as dyes, phenolic compounds, pesticides and organochlorides [31-35]. In spite of being a promising technology to decolorize molasses effluents, there are few studies concerning the oxidation of effluents from molasses-based industries.

Synthetic solutions of melanoidins prepared from glucose and glycine were decolorized to 94% after 28 h reaction time using hydrogen peroxide in a concentration of 6.72% at 37°C under alkaline pH conditions [36]. The treatment produced a reduction in the mean molecular weight of melanoidins from 5.3 to 3.5 kDa. The same procedure was followed to prepare synthetic melanoidins and to investigate their decolorization using ozone as oxidant [37]. The degree of decolorization reached 84 and 95% after 10 and 40 min reaction time. Experiments were conducted at pH about 4.4 and at low temperature, about -1°C. There was also a reduction in the mean molecular weight of melanoidins from 7 to 3 kDa. Ozone is an oxidant stronger than hydrogen peroxide and the reaction time was considerably reduced. A colored effluent from a yeast production industry was treated combining ozone with hydrogen peroxide [38]. The effluent came from two-stage aerobic biological treatment. The yields in color and COD reduction were of about 80 and 60%, respectively, after sixty minutes reaction time. Distillery wastewater from a cane molasses alcohol industry was oxidized with ozone, improving the biodegradability of the effluent and reducing color up to 80% after 40 h reaction time [39].

STUDY OF EFFLUENTS FROM AN ALCOHOLIC FERMENTATION PROCESS USING BEET MOLASSES

The main characteristics of effluents from an alcoholic fermentation factory using beet molasses as raw material are examined in more detail in this section. The suitability of chemical oxidation processes to decolorize molasses effluents was also studied.

Wastewater Characterization

Vinasse is the aqueous by-product from distillation of ethanol produced from fermentation of carbohydrates in beet molasses. Vinasse is formed by the nonfermentable soluble matter fraction of molasses as well as the nonvolatile fermentation by-products, increasing the COD content of the effluent. Some studies confirm the presence of organic compounds such as acetic acid, lactic acid, succinic acid, glycerol, inositol and glutamic acid [40]. An anaerobic digestion scheme is usually applied to treat vinasse and produce biogas. Two anaerobic reactors are usually used. First, a termophilic reactor working at temperatures about 55 °C and, then, a mesophilic reactor operating at lower temperatures, about 35°C, and treating the thermophilic effluent mixed with wastewater from the sugar factory. An aerobic treatment completes the biological treatment. A great reduction in the organic matter is achieved throughout the entire biological process, but the treated wastewater is still brown colored. There are also a high content in inorganic salts, mainly chloride, bicarbonate, sulfate and ammonia. A summary of the average composition of vinasse and the anaerobic-aerobically treated effluent, both taken from an alcohol fermentation factory, is presented here:

	Vinasse	anaerobic-aerobically treated effluent
pH	4.5-5	8.5-9
Color[a], a.u.	7-7.5	2-3
COD, mg/L	85000	4000
BOD, mg/L	76000	300
Alkalinity, mg $CaCO_3$/L	1500	9000
Sulfate, mg/L	1800	460
Chloride, mg/L	4500	800
Ammonia, mg/L	170	1000
Potassium, mg/L	7700	600
Calcium, mg/L	315	25

[a] a.u.: absorbance units measured at 475 nm

Study of Colorants in Molasses, Vinasse and Biological Process Streams

GPC Applied to Sugar Colorants

GPC has been applied to study the colorants in beet molasses processing. GPC is a suitable technique for determining the molecular weight distribution of compounds in molasses and can be used to follow colored compounds through the biological treatment of vinasse. The separation process is based on the distribution of different sized molecules between the mobile phase and the pore volume. GPC columns allow large molecules to pass trough more rapidly than smaller molecules. Therefore, GPC columns separate the compounds present in colored streams according to their molecular weight. A PDA detector allows the identification of absorbing compounds, which can be characterized by their UV-Vis spectrum. A refractive index detector in series allows the detection of colored compounds that not absorb UV-Vis radiation. By comparing the spectra of colorants in beet molasses effluents with those of synthetic colorants, previously prepared, the characterization of coloring impurities is possible.

GPC Analysis

The chromatographic system for analyzing colorants was comprised by two exclusion columns, two detectors and acquiring data software. Separation was achieved in an Ultrahydrogel™ 250 column (exclusion limit 1-80 kDa) in series with an Ultrahydrogel™ 120 column (exclusion limit 0.2-5 kDa), columns packed with hydroxilated polymetacrylated-based gel. A guard column of the same material was also used. Deionized water (Millipore Q system) was used as mobile phase at a flow rate of 0.7 ml/min at 30ºC. Identification was achieved by using a Waters 996 PDA detector coupled in series with a refractive index detector. Pullulans (non-branched polymeric sugars) and polyethylene glycol polymers were used as standards for the estimation of molecular weights. Data were processed with the software Millenium 32 Chromatography Manager. Before being analyzed by GPC, the pH of the beet molasses processing effluents was adjusted to 7. Samples were filtered through a 0,22 μm membrane before injecting onto the column.

Synthetic Colorants

Sugar beet colorants, namely melanoidins, HADPs and caramels, were synthesized according with the widely used Shore's procedure [41]. Previously to GPC analysis, samples were adjusted to pH 7 and diluted in deionized water 1:50 (v/v) in order to reduce their strong absorbance at 475 nm, the wavelength usually used to measure brown color. After a GPC analysis, melanoidins from >100 to 31 kDa, HADPs from >100 to 14 kDa and caramels of 51 kDa were found [42]. Spectra of melanoidis showed a stable absorption region between 278.8 and 326.2 nm, characteristic of melanoidins. Spectra of HADPs presented a strong absorption peak at 264.6 nm, a wavelength very close to that described previously by other authors [43,44]. Spectrum of synthesized caramels showed absorption maximum at 250.4 nm, wavelength rather different to that proposed for other researches that found sharp maxima at 220-226 nm.

Beet Molasses Colorants

A sample of beet molasses from a Spanish sugar factory was taken for the study. The molasses sample has a pH of about 6.5. Figure 1a shows the GPC chromatogram extracted at 475 nm. Molasses was divided into various peaks with different molecular weights. The first peak, appearing at 15.607 min, corresponded to a molecular weight of about 88 kDa. The last two peaks, eluted at 18.998 and 19.712 min, corresponded to lower molecular weight colorants, of about 5 and 3.5 kDa. There were also two small shoulders with molecular weights around 27-15 kDa. All compounds are responsible for the molasses color because of their absorbance at 475 nm. The area of the 88 kDa peak, the higher molecular weight colorant, represented 36% of the total area. Quantification was made by calculating the peak areas percentages at 475 nm.

All peaks presented similar absorption patterns with absorption maxima at about 270-275 nm. Spectra of 88 kDa and 3.5 kDa peaks are shown in Figures 1b and 1c, respectively, and compared with than found for synthetic melanoidin. Chromatograms were processed with the software Millenium 32 Chromatography Manager that is able to match spectra of colorants in molasses with those of the synthetic colorants.

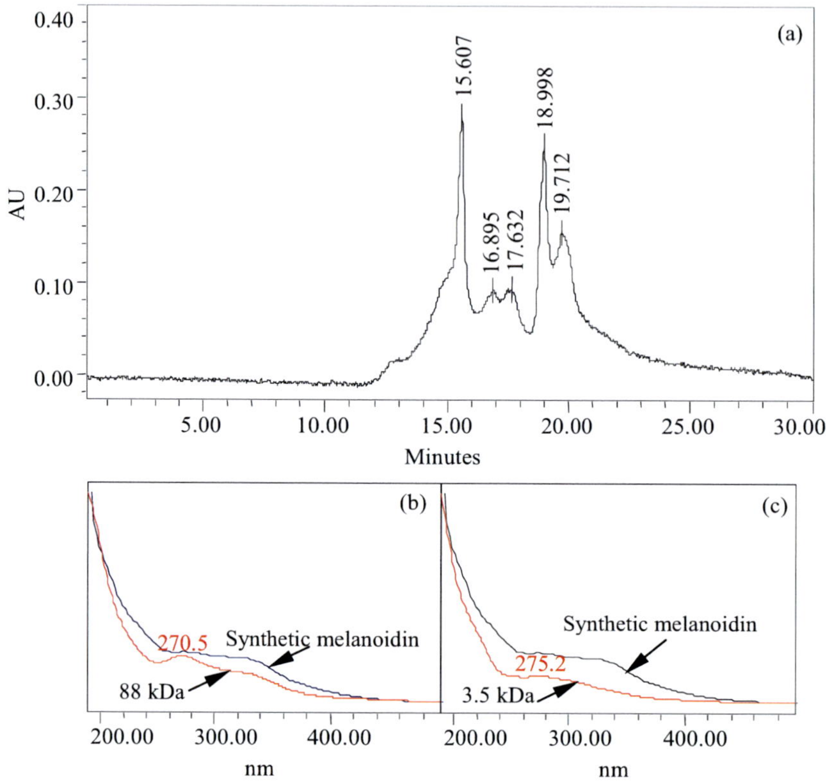

Figure 1. Beet molasses; (a) chromatogram corresponding to 475 nm; (b) spectra of synthetic melanoidin and peak with retention time of 15.607 min (88 kDa); (c) spectra of peak with retention time of 19.712 min (3.5 kDa) and synthetic melanoidin

Spectral contrast is the technique that the software uses to compare spectral shapes. PDA Match Angle is a measure of the difference in spectral shapes between an acquired spectrum and a library spectrum. Match Angle ranges from 0 to 90º. Small values indicate that spectra are similar. Large values indicate greater degrees of spectral difference. The PDA Match Angles calculated by matching the spectra of the 88 kDa and 3.5 kDa peaks with that of synthetic melanoidin were 7.0 and 14.6, respectively. The software did not find any match between the spectra of colorants in molasses and the other colorants synthesized, HADPs and caramels. The differences between the spectrum shapes may be due to the different conditions of pH and reactants between the Shore's synthesis method and the sugar crystallization stage. The pH value reached during the crystallization stage is about 6.9, much lower than the strong basic pH value, about 11, fixed during the synthetic procedure. Regarding the nature of reactants, glycine was used as reagent whereas glutamic acid and aspartic acid are the most common amino acids in sugar beet juices. Moreover, the sugar beet processing is a system of reactions and interactions much more complex than the synthetic medium.

Colorants in Vinasse and Biological Treatment Effluents

Vinasse, the effluent from the mesophilic reactor and the anaerobic-aerobically treated effluent were analyzed by GPC.

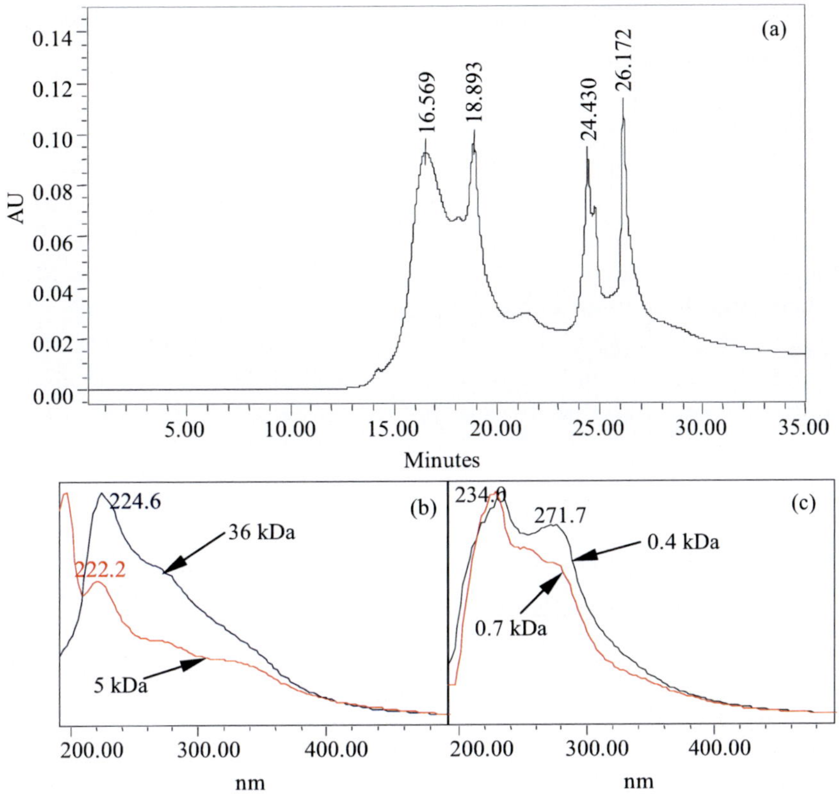

Figure 2. Beet vinasse; (a) chromatogram corresponding to 475 nm; (b) spectra of peaks with retention times of 16.569 min (36 kDa) and 18.893 min (5 kDa); (c) spectra of peaks with retention times of 24.43 min (0.7 kDa) and 26.172 min (0.4 kDa)

Figure 2a shows the chromatogram, extracted at 475 nm, corresponding to vinasse. GPC corroborated the presence of colored compounds in vinasse. There were colorants with a molecular weight of about 36 kDa (49% of the total area) corresponding to an elution time of 16.569 min, lower molecular weight colorants of about 5 kDa, with an elution time of 18.893 min, and much lower molecular weight colorants ranging from 0.7 to 0.4 kDa. Comparing the molecular weights of colorants in vinasse with those in molasses, it can be noticed that the lower molecular weight colorants in vinasse, 0.7-0.4 kDa, are likely to be formed during the alcoholic fermentation process. Spectra of the 36 and 5 kDa peaks are shown in Figure 2b. Spectrum of the 36 kDa peak showed maxima at 196.5 and 222.2 nm. PDA software matched the spectrum of the 36 kDa group with that of synthetic melanoidin, found a match value of about 6.7. Therefore, it can be supposed that melanoidins-type colorants are associated with this group of coloring components. Spectra of lower molecular weight components were not similar to any of the synthesized colorants. The 5 kDa peak showed a sharp maximum at 224.6 nm. The peaks with molecular weight about 0.7-0.4 kDa showed similar absorption patterns, with maxima at about 230 and 270 nm. These colorants are likely to be composed of mixtures of different compounds. In this case, the UV-Vis spectrum is a combination of individual absorption spectra. Moreover, melanoidins are very complex compounds, with a chemical structure with many reactive centers that make possible further reactions. The alcoholic fermentation process, taking place under acidic conditions, are likely to produce changes in the structure of melanoidins, resulting in the different spectra of the lower molecular weight colorants separated.

Figure 3a shows the chromatogram, extracted at 475 nm, corresponding to the beet vinasse after the anaerobic digestion comprising two anaerobic stages: a thermophilic followed by a mesophilic reactor. The most important compounds were eluted at times of about 16.044 and 18.812 min, corresponding to molecular weights of about 57 kDa (67%) and 5 kDa. There were also two small shoulders, corresponding to much lower molecular weights of about 1.1-0.8 kDa. The spectra of 57 and 5 kDa peaks are compared with that of synthetic melanoidins in the figures 3b and 3c, respectively. Spectra were very similar, with a stable absorption region comprised between 260-280 nm. By comparing the spectra of colorants in wastewater with that of synthetic melanoidin, the software found PDA match values of about 3.5, indicating that melanoidins are associated with these groups of colored components.

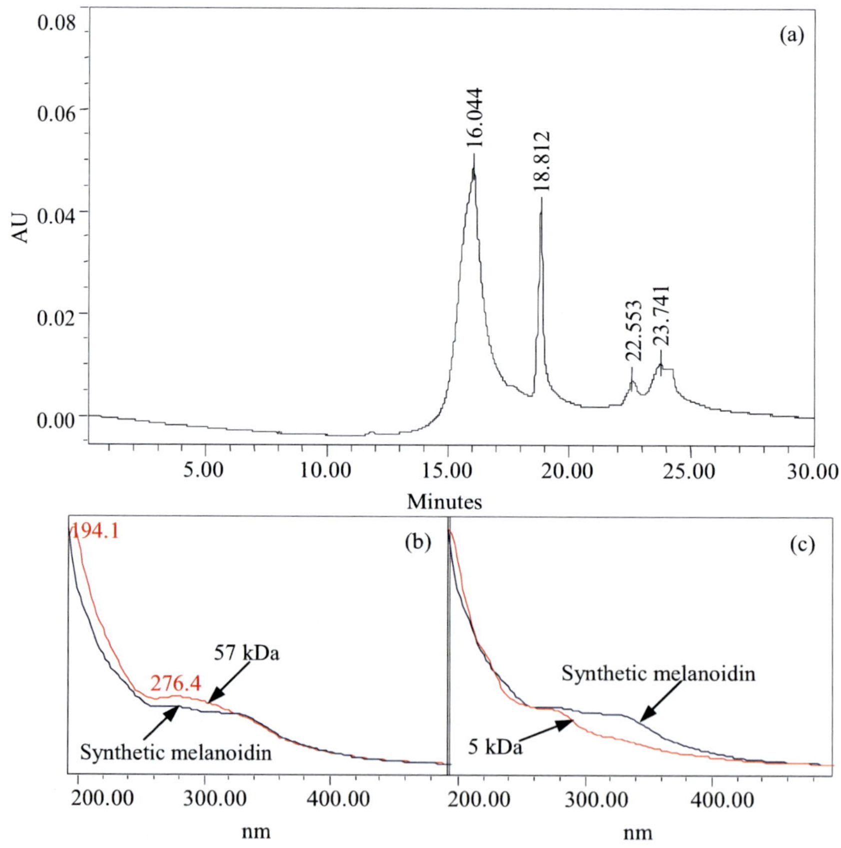

Figure 3: Mesophilic effluent; (a) chromatogram corresponding to 475 nm; (b) spectra of synthetic melanoidin and peak with retention time of 16.044 min (57 kDa); (c) spectra of peak with retention time of 18.812 min (5 kDa) and synthetic melanoidin

The molecular weight distribution corresponding to the anaerobic-aerobically treated effluent is shown in the Figure 4. As it is observed, colorants were divided in two groups. The first group comprised compounds whose molecular weight at the peak apex, elution time of 16.26 min, is about 47 kDa (31%). The second group comprised compounds with an elution time of about 20.237 min, corresponding to a molecular weight of about 3 kDa. A pronounced shoulder corresponding to a higher molecular weight component, 4 kDa, appeared on the last eluted peak. The spectra of 47 and 3 kDa peaks were compared in Figure 4b. As can be seen, both spectra evolved in the same way, with weak maxima at about 200 nm and 271.7 nm. Figure 4c compares the spectra of 47 kDa peak with that of synthetic melanoidin. The PDA match value found was 8.3. Spectral differences are likely to be due to the considerably more complex interactions of colorants during the sugar production process and the biological treatments.

Figure 5 compares the molecular weight distribution of vinasse and effluents produced throughout the biological treatment, that is, the mesophilic effluent and the anaerobic-aerobically treated effluent. As can be seen from the chromatogram, the lower molecular weight colorants in vinasse, 0.7-0.4 kDa, disappeared after the biological treatment.

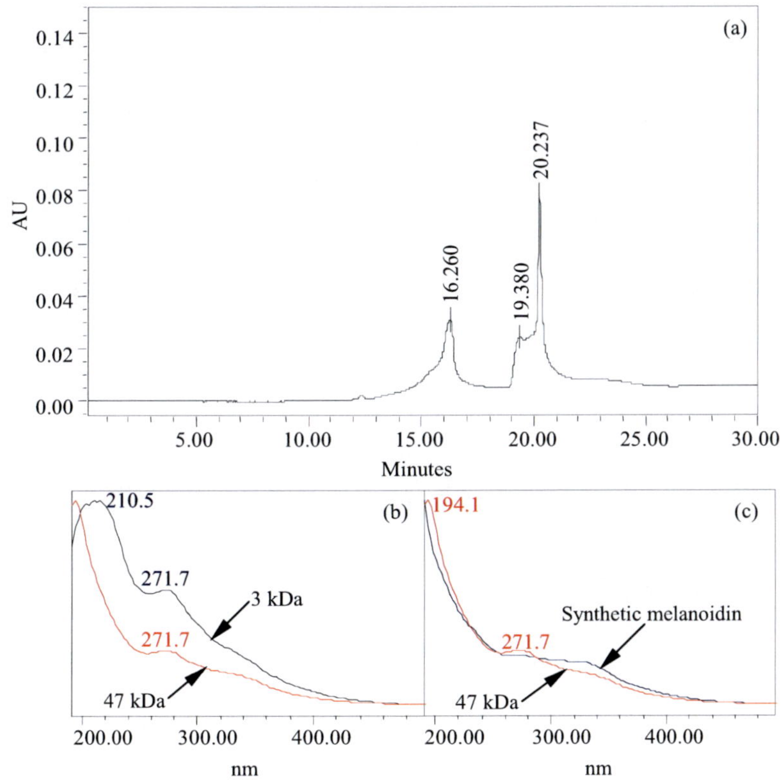

Figure 4: Anaerobic-aerobically treated effluent; (a) chromatogram corresponding to 475 nm; (b) spectra of peaks with retention time of 16.26 min (47 kDa) and 20.237 min (3 kDa); (c) comparison between spectra of 47 kDa peak and synthetic melanoidin

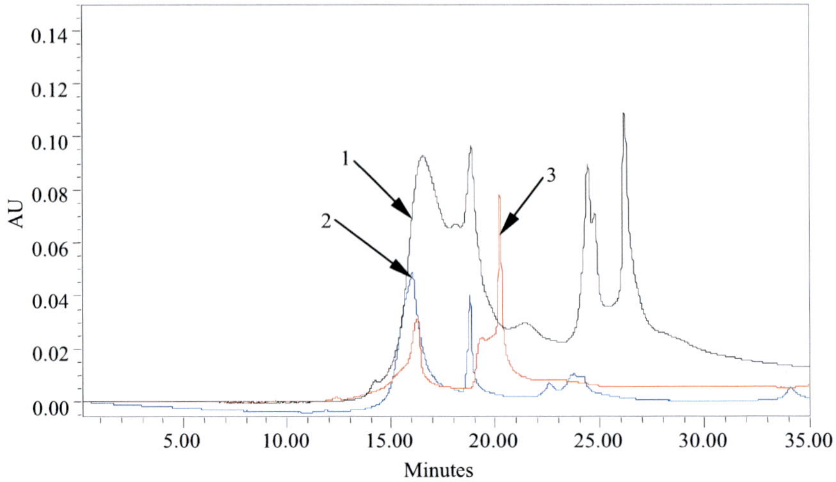

Figure 5. Molecular weight distribution throughout the biological treatment of vinasse. Chromatograms at 475 nm, corresponding to (1) vinasse, (2) mesophilic effluent and (3) anaerobic-aerobically treated effluent

Therefore, these colorants are likely to be compounds formed during the alcoholic fermentation process that can be easily degraded by the biological treatments. Moreover, the spectra of these compounds, shown in Figure 2c, indicated that were not associated with melanoidins. The absorbance of vinasse is considerably higher than those of treated effluents. However, the reduction in the absorbance was due to the dilution of the thermophilic effluent with no colored wastewater from the sugar factory. The colored compounds that remained in the effluent and, thus, were not biologically degraded, corresponded to the peaks eluted at about 16 and 19 min, with molecular weights of about 40-50 kDa and 5 kDa, respectively. The PDA spectral contrast indicated that these colored components were associated with melanoidin-type colorants, corroborating that this kind of colorants is poorly degraded by thermophilic, mesophilic and aerobic treatments.

Chemical Oxidation Process with Ozone

Chemical oxidation with ozone presents desirable properties for wastewater treatment. Ozone is a powerful oxidant, soluble in water, readily available and without by-products, other than oxygen, that need further treatments. Ozone reacts directly with an organic compound by means of an electrophilic attack, both in the Crieege mechanism and in the alternative mechanism for the oxidation of aromatic rings. In the Crieege mechanism a cycloaddition product is formed. The intermediates formed further react to yield carboxylic acids, alcohols or aldehydes. In the reaction with aromatic compounds, an addition takes place rather than a cycloaddition, and after elimination of oxygen, the phenolic reaction product is formed. The reaction follows according to the Crieege mechanism, will yield ring-opened products. Ozone is very reactive towards compounds incorporating conjugated double bonds, often associated with color, and functional groups with high electron densities. Apart from direct reactions, indirect or radical reactions also occur. The hydroxyl anion (OH^-), the hydroperoxyl anion (HOO^-), the UV radiation and certain metals, such as ferrous ions, initiate the decomposition of ozone in aqueous solutions. Decomposition reactions involve a quite complicated sequence of reactions. The most important intermediates are hydroxyl radicals ($HO^{\bullet}$), superoxide radicals ($O_2^{\bullet}$) and hydrogenperoxyl radicals ($HO_2^{\bullet}$). Secondary oxidants, mainly the non-selective hydroxyl radicals with an oxidation potential higher than that of ozone, may be responsible for a significant part of the degradation of organic compounds in the ozonation processes. Radicals are removed by termination reactions, in which two radicals combine to form a molecule.

The presence of salts affects the ozone oxidation process in two ways: physically and chemically. The presence of salts generally reduce ozone solubility but inhibit bubble coalescence, resulting in a lower bubble diameter and increasing mass transfer rates. On other hand, carbonate and bicarbonate ions are well-known species scavenging the radicals involved in the ozone decomposition. Reactions between bicarbonate ions and radicals promote termination reactions, resulting in a catalytic decrease of the hydroxyl radical concentration and a higher share of the direct oxidation pathway. The scavenging effect of other inorganic salts, such as phosphate, sulfate, chloride and nitrate, is not well established. These ions are slower in reacting with hydroxyl radicals and their scavenging effect is unlikely but it cannot be neglected.

Ozonation Experimental Set-Up

The experimental system was based on a 2 liter jacketed stirred tank reactor, equipped with a Sander 301 ozone generator, previously calibrated, producing ozone from dry air by electrical discharge. Ozone was supplied at the bottom of the reactor through a gas diffuser. Experiments were conducted at room temperature and in batch mode by sparging the ozone-containing gas continuously into the solution. The inlet gas flow rate was 200 L/h, which provided appropriate mixing conditions. Two ozone traps containing 2% potassium iodine solutions were connected in series with the reactor in order to collect ozone that did not react. Two types of wastewaters were studied: vinasse and the anaerobic-aerobically treated effluent.

Ozonation of Vinasse

The ozonation of vinasse applying an ozone dosage of 4.5 g O_3/h per liter of vinasse to be treated led to a reduction in color, measured at 475 nm after removing suspended solids, of about 76% after 30 min reaction time. A further increase in the reaction time up to 50 min also increased the color removal efficiency up to 83%. In spite of the effectiveness in reducing color, the vinasse after the oxidation treatment with ozone showed an absorbance value of about 1.3 a.u., indicating that the brown color was not sufficiently reduced. The reduction in the organic matter content was considerably lower. The COD reduction was about 10% and the percentage of TOC removed was about 9% after the 50 min reaction time. Ozonation experiments were carried out at pH 4.6, the initial pH of the vinasse.

The ozonation treatment may be able to transform the refractory compounds of high molecular weight in vinasse into simpler molecules more readily assimilated by bacteria, improving the anaerobic biodegradability of vinasse. Anaerobic degradation tests were performed to study the increase in the biodegradability of vinasse after the ozonation treatment. The test corresponding the ozone-treated vinasse produced an increase in the volume of methane of about 13% in comparison with non-ozonated vinasse. There was also an increase of about 17% in the percentage of COD removed. The methane production increased from 0.5 to 0.7 m^3 CH_4/Kg degraded COD.

The molecular weight distribution of colorants in vinasse was determined by GPC. Figure 6 compares the chromatograms corresponding to the vinasse before and after the oxidation treatment with ozone. Raw vinasse was separated into various peaks. The first peak, appearing at 16.4 min, corresponded to a molecular weight of about 40 kDa. The second peak, with a retention time of 20.12 min, corresponded to a colorant of 3 kDa molecular weight. The last peak, eluted at 23.52 min corresponded to a compound of much lower molecular weight, about 0.8 kDa. The chromatograms confirmed a sharp reduction in the absorbance of colored compounds after 50 min of oxidation treatment with ozone. The reduction in the area of the 40 kDa peak was of about 80%, lower than those achieved with the lower molecular weight colorants, 3 kDa and 0.8 kDa peaks, that were of about 92 and 86% respectively. In no case, lower molecular weight components appeared. The ozonation treatment was able to reduce the concentration of chromophore groups responsible for the brown color but it was not able to break high molecular weight colorants into smaller compounds.

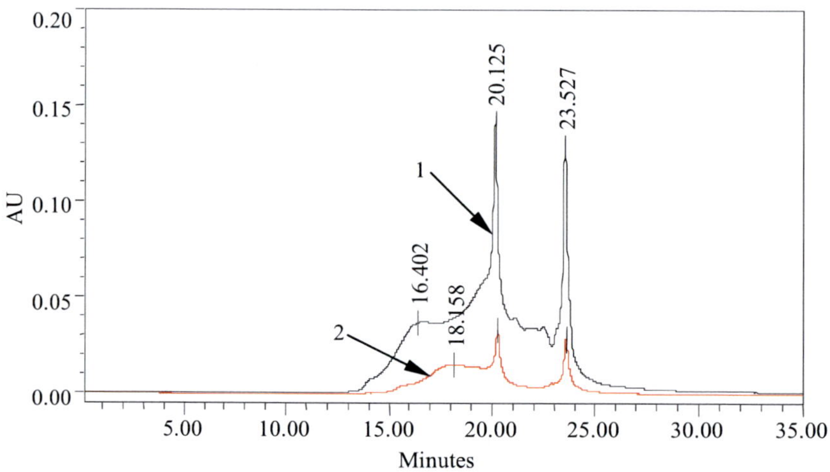

Figure 6. Molecular weight distribution (1) vinasse before ozonation; (2) vinasse after 50 min of ozonation. Chromatograms at 475 nm.

As the ozonation efficiencies are strongly dependent on the pH of the wastewater, the influence of pH on the oxidation treatment of vinasse was tested adjusting the pH at values ranging from 4 to 11. Although the results were virtually independent of the pH, the highest efficiencies according with color and organic matter removal percentages were obtained at the original pH of the vinasse.

Ozonation of Biologically Treated Effluent

The anaerobic-aerobically treated effluent was also oxidized with ozone. An ozone dosage of 2.3 g O_3/h per liter of wastewater to be treated provided a color removal efficiency of about 76% after 30 min reaction time. The color of the effluent changed considerably from dark brown to light yellow. A further increase in the reaction time did not lead to a further reduction in color. A COD removal of 13% was obtained after the same reaction time. Longer reaction times did not lead to further reductions neither in color nor COD. TOC variations were negligible throughout ozonation because mineralization of organic matter was not achieved.

Ozonation experiments were carried out at the original pH of the wastewater that was a pH about 8.7. Under basic pH conditions hydroxyl ions promotes the decomposition of ozone, producing very oxidizing and non-selective $HO^{\bullet}$ radicals. However, the biologically treated effluent contains a great concentration in inorganic salts, mainly bicarbonate ions, which have been described as strong inhibitors of reactions between organic matter and hydroxyl radicals [45]. Experiments after removing bicarbonate ions from the wastewater were carried out to perform a theoretical study on the influence of bicarbonate ions on the oxidation process. The removal of bicarbonate ions led to yields in color and COD removal higher than those achieved with the raw anaerobic-aerobic effluent. The percentages of color and COD removal increased from 71 to 84% and from 13 to 23%, respectively. There was also a slight TOC depletion of about 5%. Therefore, the bicarbonate ions in the wastewater inhibit strongly the oxidant action of $HO^{\bullet}$ radicals in the ozonation of the biologically treated effluent.

The influence of the pH on the oxidation process was also studied. The pH was varied from 5 to 11. At pH values lower than 11, the percentage of color removed after 30 min of reaction was of about 73%, independently of the pH. However, if the ozonation is carried out at pH 11 the percentage of color removed increased up to 85% and the decolorization rates also increased indicating that the radical reactions may be the main responsible for decolorization reactions. Regarding organic matter removal, there was no pH influence on the COD removal percentages, about 13-15%, and the differences were negligible regardless of the pH

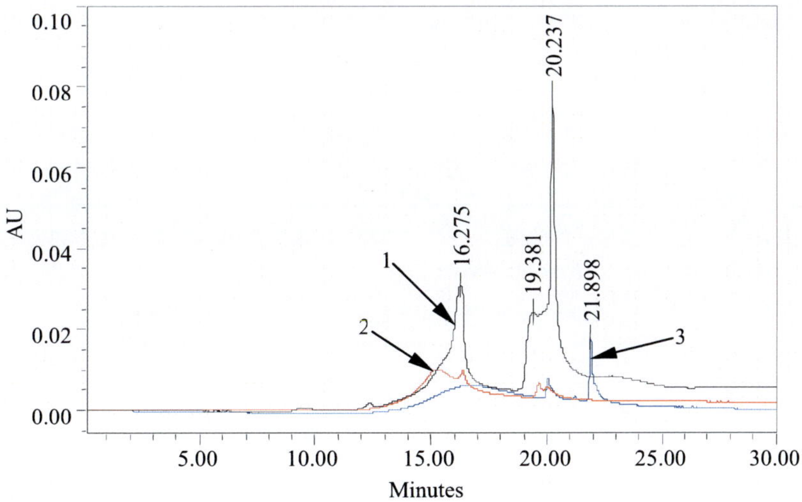

Figure 7. Molecular weight distribution (1) anaerobic-aerobically treated effluent; (2) effluent after 30 min of ozonation at pH 11; (3) effluent after 30 min of ozonation at pH 5. Chromatograms at 475 nm

Figure 7 shows the molecular weight distribution of the aerobic effluent before and after being treated with ozone at pH 11. The chromatogram, extracted at 475 nm, confirmed that there was a sharp decrease in the concentration of chromophore groups responsible for the brown wastewater color after the oxidation treatment. The reduction in the area of the 47 kDa peak, elution time of 16.275 min, was of about 75%. The reduction in the area of colorants with molecular weight of 3-4 kDa, was much higher, of about 95%, indicating that the lower molecular weight colorants are more efficiently oxidized by ozone.

The combination of ozone with hydrogen peroxide aims to increase the production of $HO^{\bullet}$ radicals from ozone decomposition. The rate of hydroxyl radicals production is considerably much higher than that achieved when decomposition is initiated by hydroxyl ions. Different concentrations of hydrogen peroxide, ranging from 2 to 100 mM, were initially added to the wastewater. Immediately after, ozone was continuously fed up during 30 min. A low concentration of hydrogen peroxide, 2 mM, was enough to enhance efficiencies in relation to those achieved throughout single ozonation. Color removal increased from 76 to 89% after 30 min reaction time, whereas the COD removed increased from 13 to 23% after the same reaction time. TOC values remained constant. A further increase in the initial hydrogen peroxide concentration did not lead to an increase in the removal efficiencies.

Other Oxidation Processes

The ability of hydrogen peroxide to decolorize vinasse was also tested. Hydrogen peroxide in a concentration of 15 mM was initially added to vinasse. However color remained constant after 24 h reaction time. The dosage of peroxide was two-fold increased after 24 h, but color was not reduced after 48 h reaction time. Therefore oxidation with hydrogen peroxide alone was not effective in degrading colorants in vinasse.

With reference to the anaerobic-aerobically treated effluent, chemical oxidation using single hydrogen peroxide and Fenton's reagent were also studied in addition to ozonation. Hydrogen peroxide has an oxidation potential lower than that of ozone, but can be combined with iron at acidic pH values. Iron can promote the hydrogen peroxide decomposition to form hydroxyl radicals, which are stronger oxidants. Regarding single hydrogen peroxide oxidation, an initial concentration of 15 mM did not lead to changes in neither color nor COD after 40 h reaction time. With regard Fenton's experiments, the initial hydrogen peroxide concentration was also 15 mM and iron was added in a molar ratio H_2O_2/Fe (II) of about 24:1. However, Fenton´s reagent was not effective in reducing neither color nor organic matter. An increase in the concentrations of both hydrogen peroxide, from 15 to 30 mM, and iron but maintaining the molar ratio H_2O_2/Fe, resulted in the removal of a high percentage of color and organic matter due to coagulation and flocculation processes, promoted by the presence of the high concentration of iron and the acidic pH value, about 5, required by Fenton´s experiments. The percentage of color removal was of about 75% and the efficiency in removing TOC was around 46%.

Therefore, this study confirmed that the oxidation process with ozone is an appropriate technology for reducing color from molasses effluents. Regarding raw vinasse, a color removal higher than 80% was achieved together with a slight increase in the anaerobic biodegradability. The ozonation treatment applied to the anaerobic-aerobically treated effluent yielded a color removal percentage of about 76% and the brown color almost disappeared, avoiding the detrimental impact of the effluent on environment. A low concentration of hydrogen peroxide, 2 mM, led to an improvement in the efficiencies reached throughout single ozonation runs. The percentages of organic matter removal were, in any case, considerable lower that those achieved in color.

REFERENCES

[1] Schiweck, H. (1994). Composition of sugar beet molasses. *Zuckerindustrie*, 119, 272-282.

[2] Higginbotham, J.D., and McCarthy, J. (1998) Quality and storage of molasses. In P.W. van der Poel, H. Schiweck, and T. Schwartz (Eds.), *Sugar Technoloy: Beet and Cane Sugar Manufacture* (973-992). Berlin, Verlag Dr. Albert Bartens KG.

[3] González, G., and García, M.T. (1996). Ammonia elimination from beet sugar factory condensate streams by a stripping -reabsorption system. *Zuckerindustrie*, 121, 721-726.

[4] Bruhns, G., Riffer, R., van Bekkum, H., Lammers, H., Schiweck, H., Heitz, F., and Mauch, W. (1998). Sugar. In P.W. van der Poel, H. Schiweck, and T. Schwartz (Eds.), *Sugar Technoloy: Beet and Cane Sugar Manufacture* (37-114). Berlin, Verlag Dr. Albert Bartens KG.

[5] Perschak, F. (1998) Molasses desugarizing. In P.W. van der Poel, H. Schiweck, and T. Schwartz (Eds.), *Sugar Technoloy: Beet and Cane Sugar Manufacture* (939-950). Berlin, Verlag Dr. Albert Bartens KG.

[6] Decloux, M., and Bories, A. (2002). Stillage treatment in the French alcohol fermentation industry. *International Sugar Journal*, 104, 509-517.

[7] Madejón, E., Díaz, M.J., López, R., and Cabrera, F. (2001). Co-composting of sugarbeet vinasse: influence of the organic matter nature of the bulking agents used. *Bioresource Technology,* 76, 275-278.

[8] Madejón, E., Lopez, R., Murillo, J.M., and Cabrera, F. (2001). Agricultural use of three (sugar beet) vinasse compost: effect on crops and chemical properties of a Cambisol soil in the Guadalquivir river valley (south Spain). *Agriculture Ecosystems and Environment*, 84, 55-65.

[9] Vaccari, G., Tamburini, E., Sgualdino, G., Urbaniec, K., and Klemes, J. (2003). Overiew of the environmental problems in beet sugar processing: possible solutions. *Journal of cleaner production,* in press.

[10] Wilkie, A.C., Riedesel, K.J., and Owens, J.M. (2000). Stillage characterization and anaerobic treatment of ethanol stillage from conventional and cellulosic feedstocks. *Biomass and bioenergy*, 19, 63-102.

[11] Cämmerer, B., and Kroh, L.W. (1995). Investigations of the influence of reactions conditions on the elementary composition of melanoidins. *Food Chemistry*, 53, 55-59.

[12] Heitz, F. (1995). Les colorants en sucrerie. *Actualités techniques et industrielles*, 538-544.

[13] Martins, S.I.F.S., Jongen, W.M.F., and Van Boekel, M.A.J.S. (2001). A rewiew of Maillard reaction in food and implications to kinetic modelling. *Trends in Food Science and Technology*, 11, 364-373.

[14] Reinefeld, E., Bliesener, K.M., Brandes, E., and Borras, V. (1982). Contribution to the knowledge of color formation in sugar beet juices. *La sucrerie belge*, 101, 333-345.

[15] De Bruijn, J.M. (1986). Monosaccharides in alkaline medium: isomerization, degradation and oligomerization. *Ph.D. Thesis*, University of Technology, Delf.

[16] Kroh, L.W. (1994). Caramelisation in food and beverages. *Food Chemistry*, 51, 373-379.

[17] Goto, M., Nada, T., Ogata, A., Kodama, A, and Hirose, T. (1998). Supercritical water oxidation for the destruction of municipal excess sludge and alcohol distillery wastewater of molasses. *Journal of Supercritical Fluids*, 13, 277-282.

[18] Mutlu, S.H., Yetis, U., Gurkan, T., and Yilmaz L. (2002). Decolorization of wastewater of a baker´s yeast plant by membrane processes. *Water Research*, 36, 609-616.

[19] González Benito, G., Peña Miranda, M., García Cubero, M.T., and Urueña Alonso M.A. (1999). Decolorization of molasses effluents by coagulation-flocculation process. *Zuckerindustrie*, 124, 406-410.

[20] Migo, V.P., Matsumura, M., Del Rosario, E.J., and Kataoka, H. (1993). Decolorization of molasses wastewater using an inorganic floculant. J. *Ferment. Bioeng.*, 75, 438-442.

[21] Dahiya, J., Singh, D., and Nigam, P. (2001). Decolourisation of molasses wastewater by cells of Pseudomonas fluorescens immobilised on porous cellulose carrier. *Bioresource Technology*, 78, 111-114.

[22] Nakajima-Kambe, T., Shimomura, M., Nomura, N., and Chanpornpong, T., and Nakahara, T. (1999). Decolorization of molasses wastewater by Bacillus sp. under thermophilic and anaerobic conditions. *Journal of Bioscience and Bioengineering*, 87, 119-121.

[23] Sirianuntapiboon, S., Phothilangka, P., and Ohmomo, S. (2004). Decolorization of molasses wastewater by a strain NoBP103 of acetogenic bacteria. *Bioresource Technology,* 92, 31-39.

[24] Sirianuntapiboon, S., Zohsalam, P., and Ohmomo S. (2004). Decolorization of molasses wastewater by a Citeromyces sp. WR-43-6. *Process Biochemistry*, 39, 917-924.

[25] Sirianuntapiboon, S., Sihanonth, P., Somchai, P., Atthasampunna, P., and Hayaschida, S. (1991). Mechanism for melanoidin decolorization by Rhizoctonia sp. D-90. *Biosci. Biotech. Biochem.*, 59, 1185-1189.

[26] Jiménez, A.M., Borja, R., and Martín, A. (2003). Aerobic-anaerobic biodegradation of beet molasses alcoholic fermentation wastewater. *Process Biochemistry*, 38, 1275-1284.

[27] González Benito, G., Peña Miranda, M., Rioja Ribón, R., Coca Sanz, M., and García Cubero, M.T. (2001). Development of biological treatment process with Trametes versicolor to decolorize wastewater from alcoholic molasses fermentation. *In Proceedings International Water Conference*, Porto (Portugal), 259-263.

[28] Kumar, V., Wati, L., Nigam, P., Banat, I.M., Yadav, B.S., and Singh, D. (1998). Decolorization and biodegradation of anaerobically digested sugarcane molasses spent wash effluent from biomethanation plants by white rot fungi. *Process Biochemistry*, 33, 83-88.

[29] Guimaraes, C., Bento, L.S.M., and Mota, M. (1999). Biodegradation of colorants in refinery effluents. Potential use of the fungus Phanerochaete chrysosporium. *International Sugar Journal*, 101, 246-251.

[30] Miyata, N., Mori, T., Iwahori, K., and Fujita, M. (2002). Microbial decolorization of melanoidin-containing wastewaters: combined use of activated sludge and the fungus Coriolus hirsutus. *Journal of Bioscience and Bioengineering*, 89, 145-150.

[31] Ormad, P., Cortés, S., Puig, A., and Ovelleiro, J.L. (1997). Degradation of organochloride compounds by O_3 and O_3/H_2O_2. *Water Research*, 31, 2387-2391.

[32] Chiron, S., Fernandez, A., Rodriguez, A., and García, E. (2000). Pesticide chemical oxidation: State of the art. *Water Research*, 34, 366-377.

[33] Beltran-Heredia, J., Torregrosa, J., Joaquin, R., Dominguez, J.R., and Peres, J.A. (2001). Kinetics of the reaction between ozone and phenolics acids presents in agro-industrial wastewaters. *Water Research*, 35, 1077-1085.

[34] Koch, M., Yediler, A., Lienert, D., Insel, G., and Kettrup, A. (2002). Ozonation of hydrolyzed azo dye reactive yellow 84 (CI). *Chemosphere*, 46, 109-113.

[35] Wu, J.J., and Masten, S.J. (2002). Oxidation kinetics of phenolic and indolic compounds by ozone: applications to synthetic and real swine manure slurry. *Water Research*, 36, 1513-1526.

[36] Hayase, F., Kim, S.B., and Kato, H. (1984). Decolorization and degradation products of the melanoidins by hydrogen peroxide. *Agricultural and Biological Chemistry*, 48, 2711-2717.

[37] Kim, S.B., Hayase, F., and Kato, H. (1985). Decolorization and degradation products of melanoidins on ozonolysis. *Agricultural and Biological Chemistry*, 49, 785-792.

[38] Filipovic-Kovacevic, Z., and Sipos, L. (1995). Decolorization of yeast production industry wastewater by ozone. *J. Environ. Sci. Health.*, A30, 1515-1522.

[39] Alfafara, C.G., Migo, V.P., Amarante, J.A., Dallo, R.F., and Matsumura, M. (2000). Ozone treatment of distillery slop waste. *Water Science and Technology*, 42, 193-198.

[40] Malnou, M., Huyard, A., and Foup, G.M. (1987). High load process using yeast for vinasse of beet molasses treatment. *Water Science and Technology* 19, 11-21.

[41] Shore, M., Broughton, N.W., Dutton, J.V., and Sissons, A. (1984). Factors affecting white sugar colour. *Sugar Technology Reviews*, 12, 1-99.

[42] Coca, M., García, M.T., González, G., Peña, M., and García, J.A. (2004). Study of colored components formed in sugar beet processing. *Food Chemistry*, 86, 421-433.

[43] Guimaraes, C., Bento, L.S.M., and Mota, M. (1996). A study of sugar colorants through ion exchange and salt regeneration. *International Sugar Journal*, 98, 584-587.

[44] Mersad, A., Lewadowski, R., Heyn, B. and Decloux, M. (2001). Formation des colorants de fabrication en sucrerie: reconstitution au laboratorie et analyses par spectrométrie. *Industries Alimentaries et Agricoles*, 118, 42-51.

[45] Glaze, W.L., and Kang, J.W. (1989). Advanced oxidation processes. Description of a kinetic model for the oxidation of hazardous materials in aqueous media with ozone and hydrogen peroxide. *Industrial and Engineering Chemical Research*, 28, 1573-1580.\\

In: Water Pollution: New Research
Editor: A. R. Burk, pp. 55-76

ISBN: 1-59454-393-3

Chapter 3

ENHANCED LOSSES OF PHOSPHORUS IN MOLE-TILE DRAINAGE WATER FOLLOWING SHORT-TERM APPLICATIONS OF DAIRY EFFLUENT TO PASTURE

***R. W. McDowell**[*,1], **R. M. Monaghan**[1], **L. C. Smith**[2], **G. F. Koopmans**[3] **and I. Stewart**[4]*

[1]AgResearch Ltd, Invermay Agricultural Centre, Private Bag 50034, Mosgiel, New Zealand

[2]AgResearch Woodlands, RD 1, Invercargill, New Zealand

[3]Alterra, Wageningen University and Research Centre (WUR), Wageningen, The Netherlands

[4]Department of Chemistry, University of Otago, Dunedin, New Zealand

ABSTRACT

The loss of phosphorus (P) in drainage waters can cause eutrophication and impairment of surface water quality. Dissolved organic P (DOP) is available to algae and also leaches farther down the soil profile than dissolved inorganic P, especially in soils that have received long-term applications of manure or effluent. Best management stipulates that P fertiliser applications should account for P added in effluent to not overload soil with P and risk P loss to surface waters. Hence, the aim of this study was to compare the form and losses of P in mole-tile drainage waters from a plot receiving dairy effluent to one that received P at the same rate but as superphosphate. Data over 3 years indicated that much more P was lost from the effluent-irrigated plot (mean TP load = 0.41 kg ha^{-1}) compared to the non-effluent plot (mean TP load = 0.20 kg ha^{-1}), especially via incidental transfers (events coinciding with effluent application or within a week of cattle grazing, up to 56% in the first year). Losses of DOP were, on average, 3.7 times greater in the effluent irrigated compared to the non-effluent plot. When incidental transfers were excluded, a significant relationship still occurred between the difference in DOP concentrations between plots, and event number, since irrigation began, indicating that DOP loss is enhanced by short-term effluent applications. Analysis of soil samples taken

[*] Corresponding author (richard.mcdowell@agresearch.co.nz)

from each plot indicated that more P was held in the Ca-P (due to Ca application in effluent), water soluble P and NaOH organic P fractions of the effluent irrigated soil than the non-effluent soil. Analysis of raw effluent and effluent and non-effluent treated soils by ^{31}P nuclear magnetic resonance spectroscopy indicated that much organic P (present in the NaOH organic P fraction) was phytate (450 and 386 mg kg^{-1} in the 0-2 cm depth of effluent and non-effluent treated soils, respectively). The enhanced loss of P in drainage waters from effluent irrigated soil was attributed to an increase in soil pH, and competition of P sorption sites by phytate, organic compounds and P from P-saturated colloids in the effluent. Since DOP is poorly held by soil compared to orthophosphate, we concluded that any mechanisms influencing P desorption will influence DOP loss more than DRP, especially in terms of P desorption. This highlights the need to account for DOP loss and bioavailability in soils receiving effluent even in the short-term.

INTRODUCTION

The fate of organic phosphorus (P) in soils has been the focus of a large number of studies. Many organic P species are available to aquatic organisms and may play a role in eutrophication of surface waters (e.g., Whitton et al., 1991; Turner et al., 2002). Some studies have highlighted the enhanced mobility of some organic P species over orthophosphate (e.g., Frossard et al., 1989; Hoffman and Rolston, 1980; Leytem et al., 2002). In soils receiving long-term additions of animal effluent which can contain much organic P compared to mineral fertilisers, P has been found to leach down the soil profile. For instance, Chardon et al. (1997) found that the application of pig slurry to a sandy soil in the Netherlands had enriched P concentrations down the soil profile, and that with increasing depth, dissolved organic P (DOP) was by far the dominant P fraction. Similarly, Eghball et al. (1996) compared the long-term (since 1953) application of cattle manure to a similar application rate of triple superphosphate and found that while P was enriched at depth (0.9 m) in both treatments, P moved further down, and in greater concentrations, in the manure treatment. This was again attributed to the enhanced mobility of organic P species.

In artificially drained soils, evidence has indicated that dissolved P, and to some extent colloidal associated P, can move rapidly down macropores greater than 1 mm in diameter, and appear in drainage waters (Goehring et al., 2001). The presence of macropores gives any liquid animal slurry a preferential flow pathway in which to flow, should wet conditions preclude the adsorption of the slurry by the soil matrix. In a study of the application of fresh cattle faeces to undisturbed soil columns, Jensen et al. (2000) found that earthworm burrows formed preferential flow pathways, which constituted the main mechanism of P loss under saturated conditions. However, when conditions were switched to unsaturated flow the loss of dissolved inorganic P temporarily stopped while the concentration of DOP remained unchanged. While Goehring et al. (2001) stated that soluble P loss in macropores was attributed to negligible P sorption to pore walls, the study of Jensen et al. (2000) supports earlier organic P sorption experiments that demonstrate the loss of DOP over dissolved inorganic P. In New Zealand and many other countries where effluent is used to irrigate pastures, good management dictates that the application of P fertilizer is adjusted to account for P applied in effluent to not overload the soil with P and promote P loss. Consequently, this raises the question “will effluent applications to drained pastures promote leaching and the loss of P in drainage waters, especially in DOP forms”. Our objective was to test this hypothesis on a 3-year mole-tile drained field trial that compared the amounts and forms of P lost from one plot receiving regular applications of P via dairy-shed effluent to one that received P as triple superphosphate at the same P rate.

MATERIALS AND METHODS

Site Description

The experimental site was located 10 km north-west of Tapanui, West Otago in the South Island of New Zealand (NZ map grid 2210200E, 5472660N). In October 1999, two plots (27m wide by 35m long) were hydrologically isolated from one another and tile drains installed at a depth of 75cm. Two Aquaflex tapes (Streat instruments, Christchurch, NZ) were installed at 5 to 20 cm depth to continuously measure soil moisture levels and soil moisture deficit. In November, mole drains were pulled in each plot at a depth of 45cm to connect with the tile line. Prior to installation, the site had been in pasture for at least 12 years and had a total N and C concentration in the top 7.5 cm of 4.3 and 51 g kg^{-1}, respectively. The soil is a Waikoikoi silt loam (NZ Classification: Mottled Fragic Pallic soil, USDA Taxonomy: Typic Hapludalf), which covers much of the West Otago region, but requires drainage if used for intensive agriculture. The site has an average annual rainfall of 750 mm with approximately 120 mm of drainage.

Beginning with the first drainage event of 2001 in May, drainage water volumes were continuously recorded throughout 3 years of study using 3 L tipping buckets connected to a Campbell Scientific CR10 datalogger. Samples of drainage water from each event were collected on a flow-proportional basis for analysis of P, sediment and faecal indicator bacteria (*E. coli*). Plots were denoted as either effluent or non-effluent irrigated, with effluent from a nearby milking shed being applied to the effluent plot on seven occasions during the trial via a rotating twin-gun travelling irrigator. During this time, applications of P fertiliser in early summer were adjusted to take into account the quantity of total P applied in the effluent. Hence, in 2001, the effluent plot and non-effluent plot received 48 and 50 kg P ha^{-1}, respectively, while in 2002 this was 40 and 50 kg P ha^{-1}, and in 2003, 7 and 0 kg P ha^{-1} was applied. For N, in 2001, the effluent plot received 159 kg N ha^{-1}, while the non-effluent plot received 30 kg N ha^{-1}. In 2002, the quantities were 204 and 39, respectively, while in 2003, 31 and 39 kg N ha^{-1} was applied. Previous work in the region has shown that different N applications do not significantly influence the amount of DRP or DOP lost in leachate (McDowell and Monaghan, 2002). During 2001, effluent also supplied 191 kg K ha^{-1}, 2292 kg Ca ha^{-1}, and 41 kg Mg ha^{-1}. The corresponding values for 2002 and 2003 were 211 kg K ha^{-1}, 78 kg Ca ha^{-1}, and 20 kg Mg ha^{-1}, and 11 kg K ha^{-1}, 25 kg Ca ha^{-1}, and 4 kg Mg ha^{-1}, respectively. With the exception of winter months (June – July), regular rotational grazing (every 20-30 days) of both plots occurred simultaneously during the trial at an annual stocking rate equivalent to 3.0 cows ha^{-1}.

Drainage Water Analyses

A 1 L subsample of each event was taken for analysis. Turbidity was measured as standard nephelometric turbidity units (NTU) and suspended solids (SS) by drying the filtered (< 1 μm) residue of a known volume of water. *Escherichia coli* was measured as the preferred faecal indicator bacteria for freshwater in New Zealand (MfE, 2002). Samples (flow and dung) were enumerated using the Colilert® media and the Quanti-Tray® enumeration system (IDEXX Laboratories, Maine, USA). For P, samples were filtered (< 0.45 μm) and analyzed for dissolved reactive P (DRP) immediately and total P (TP) and total dissolved P (TDP) measured after $K_2S_2O_8$ digestion (0.15 g $K_2S_2O_8$ in 1 mL 0.5M H_2SO_4 added to sample and heated at 150°C for 2 h). All P determinations were made using the colorimetric method of Watanabe and Olsen (1965). An unfiltered sample was also digested and TP measured within 7 days. Fractions defined as dissolved unreactive (largely organic) P (DOP) and particulate P (PP) were determined as TDP less DRP and TP less TDP, respectively.

Soil Analyses

In late November 2003, soils were sampled from the 0-2, 0-7.5, 7.5-15 and 15-45 cm depths at six equally spaced locations along a diagonal transect of each plot. These samples were air-dried, sieved (< 2 mm) and kept separate from one another during most analyses. The exception was for NaOH-EDTA extraction, and the generation of P sorption isotherms, whereby soils from the same depth and plot were bulked together in equal proportions. Soils were analyzed for bicarbonate extractable P (Olsen P) using the method of Olsen et al. (1954), exchangeable bases using ammonium acetate (Sumner and Miller, 1996), and organic C and organic matter by a LECO© C/N analyzer.

For P sorption isotherms, bulked soils (1 g) from each depth and treatment were shaken with 20 mL of solutions containing either 0, 1, 2.5, 10, 20 or 50 mg P L^{-1} as KH_2PO_4 for 16 h, centrifuged (4000 × *g*) and the supernatant analyzed for P remaining in solution. Data for the concentration of P sorbed (*Q*, mg kg^{-1}) versus the concentration of P remaining in solution (*C*, mg L^{-1}) were fitted by the Freundlich equation:

$$Q = \alpha C^{1/\beta}$$

whereby, α and β are constants that relate to, but do not quantify, the total quantity of P sorbed and the strength of sorption, respectively.

P Fractionation

To elucidate the chemical speciation of P within effluent and non-effluent treated soils the detailed chemical fractionation scheme of Golterman (1996) and Goletman et al. (1998) was used. Unlike many fractionation schemes used for soil P analysis, the fractions within the scheme of Golterman, along with those of the SEDEX method (Ruttenberg, 1992), have been quantified against known P compounds. Briefly, for each soil, one g is sequentially extracted via shaking with 30 mL of deionised water (2 h), 0.05M Ca-EDTA (+ 1% Na-dithionite, pH 7.8), 0.1M Na-EDTA (pH 4.5), 0.5M H_2SO_4, cold 0.5M Trichloroacetic acid (TCA; 0°C, 4 h), hot 0.5M TCA (95°C, 30 min.) and finally 2M NaOH (90°C, 1 h) before the remaining P is removed by persulphate digestion (0.15 g $K_2S_2O_8$ in 1 mL 0.5M H_2SO_4, 150°C for 2 h). These fractions remove, in sequential order, water soluble or soil solution P (H_2O-P), Fe associated P, Ca associated P, acid soluble organic P (ASOP), sugar bound P after digestion by $K_2S_2O_8$ (cold TCA), nucleic P and polyphosphate after digestion by $K_2S_2O_8$ (hot TCA, Golterman, 1960), humic bound P and phytate (NaOH P_i and NaOH P_o), and residual P. Following extraction each soil-suspension was centrifuged (4000 × *g*) for 10 min, and an aliquot taken for P determination. For Ca- and Na-EDTA extracts a maximum of 2 mLs could be used before EDTA interferes with the Mo-P colorimetric reaction. For H_2O-P, Fe-P, Ca-P, and NaOH fractions an organic P fraction was defined as the difference between P detectable before and after digestion by $K_2S_2O_8$.

^{31}P NMR

Further analysis of soil and effluent P forms was conducted using ^{31}P nuclear magnetic resonance spectroscopy (^{31}P NMR). Bulked soil samples (5 g) from each depth interval or 100 mL of fresh effluent were shaken with 100 mLs of 0.25M NaOH + 0.05M EDTA (Na form) for 16 h, centrifuged (4000 × *g*) and then analyzed for molybdate reactive P after digestion by $K_2S_2O_8$. Each extract was then frozen and freeze-dried. Approximately 0.5 g of the freeze-dried soil extract material was then re-dissolved in 0.8 mL of 10M NaOH and 1 mL of D_2O (for signal lock) and transferred into 5-mm NMR tubes. The pH of the suspension was > 12. Solution ^{31}P NMR spectra for soils were obtained using a Bruker (Rheinstetten, Germany) DPX 300 spectrophotometer operating at 121.49 MHz at 20°C. For each sample, 1024 scans were accumulated using a pulse angle of 90°, a pulse delay of 2 s, and an acquisition time of 0.67 s. Chemical shifts (δ ppm) were recorded relative to an external 0.98mM methylenediphosphonic acid standard (MDP; 98%, trisodium salt tetrahydrate) measured simultaneously in a capillary tube.

For the effluent sample, spectra were obtained using a Varian 500MHz Inova NMR spectrometer with a 51 mm standard Oxford superconducting magnet, FTS temperature controller, dual fullband channels, one low band decoupler channel and a 28 shim set. A 5mm Varian z-axis PFG Direct detection probe was used for all the samples. The Sun Ultra 10 workstation uses Solaris 8 OS and Varian VNMR 6.1C NMR software. The effluent sample was prepared to pH > 12 by taking 0.2g of the dried effluent extract, adding 600 µL of D_2O and 100 µL of 10M NaOH. Samples were ultrasonicated (Crest model 175T) for 3 minutes, equilibrated for 20 minutes then centrifuged (Qualitron 6 place mini-centrifuge) for 5 minutes. The supernatent was transferred to a NMR tube and ^{31}P NMR spectra obtained at 202.298 MHz at 20°C. Accumulation of data for each sample was halted when a sufficient signal to noise ratio was obtained after 3 hours. Scans were accumulated using a pulse angle of 45°, a pulse delay of 1 s, and an acquisition time of 1.99 s with 64K data points. Chemical shifts were recorded relative to an external phosphoric acid standard (δ=0 ppm) in a capillary tube.

Spectra were deconvoluted using a Lorentzian line shape of 10 Hz and measured using Mestre-C software (Gómez and López, 2004). Organic P compounds within the spectra were determined semi-quantitatively using the peak assignments of Cade-Menun and Preston (1996) and Turner et al. (2003), the percentage spectral area occupied by each compound, and the P concentration of the corresponding NaOH-EDTA extract.

Sorption-Desorption of Known Compounds

In order to determine the influence of organic P species and dissolved and particulate compounds in effluent on P sorption and desorption, soils from each depth and plot were treated and incubated with a range of solutions. These treatments were:

1. Control: deionised water, no P added.
2. Inorganic P (P_i) at the DRP of the effluent sample (16.7 mg P L^{-1}).
3. Inorganic P at the TDP of the effluent sample (19 mg P L^{-1}).
4. Synthetic effluent (DRP + DOP): inorganic P (16.7 mg P L^{-1} 88% P_i) + 1.3% glucose-6-phosphate + 0.4% polyphosphate + 0.4% polyphosphate + 11.7% phytate (as determined by the relative areas occupied by each class of compound in the ^{31}P NMR spectra of a fresh effluent sample, see Fig. 1).
5. Filtered effluent (< 0.45 µm; TDP, 19 mg P L^{-1}).
6. Whole effluent (TP, 60 mg P L^{-1}).

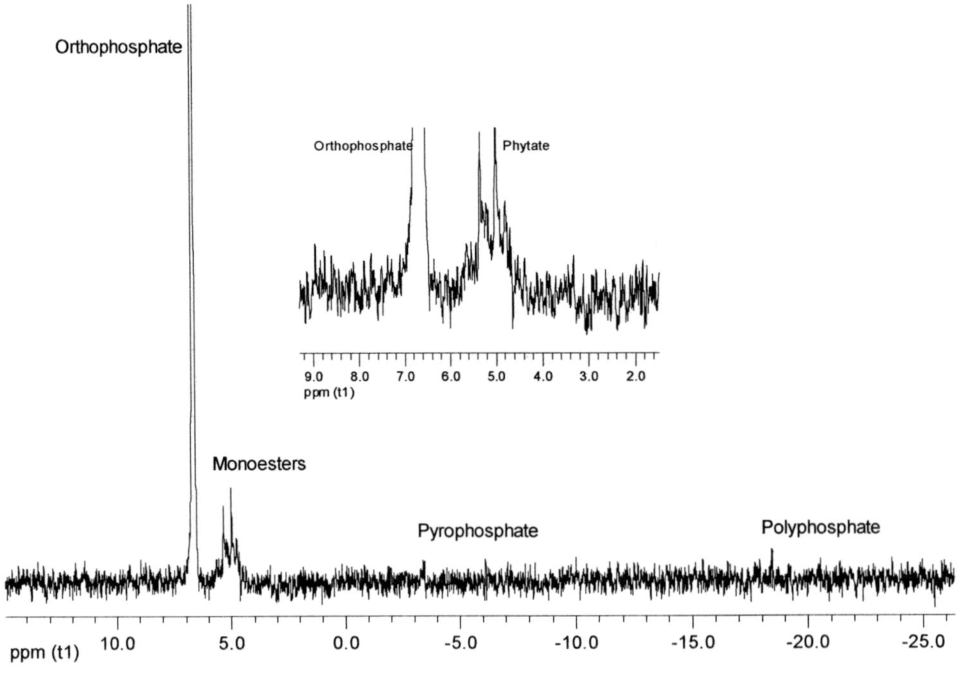

Fig. 1. ^{31}P NMR spectra of a fresh effluent sample. The expansion shows the monoester range (identified mainly as phytate)

Twenty mL of each solution, equivalent to the mean volume and P applied during an effluent application to a depth of 1 cm, was added to each soil sample, shaken for 16 h, centrifuged (4000 × *g*), P determined in the supernatant and P sorbed calculated by difference from P added. After sorption, the supernatant was discarded and 20 mL of deionised water added for desorption of P. After shaking for 2 h, the suspension was centrifuged (4000 × *g*) and DRP and DOP determined in the supernatant.

All comparisons were made by first subtracting the control. Hence, the influence of P_i is given by 3-1, while the influence of synthetic DOP compounds is given by a comparison of (4-1) – (2-1) versus (3-1) – (2-1). The influence of effluent materials other than P_i is likely from a comparison of (4-1) and (5-1), while the influence of PP and colloidal materials is given by difference of (5-1) and (6-1).

Statistical Analyses

All summary data (means, standard errors of the mean, and least significant differences at the $P < 0.05$ level) for soil analyses, sequential P fractions, and organic P compounds identified by ^{31}P NMR were assessed using SPSS v10.0 (SPSS Inc, 1999). Curve-fits for P sorption isotherm data were also fitted to the Freundlich equation using SPSS v10.0. All curve fits were significant at the 5% level or better and yielded a coefficient of determination in excess of 0.96.

Results and Discussion

Drainage Losses

Analysis of flow data for this site by Monaghan et al. (2002) determined that averaged across both plots, *c.* 96% of surplus rainfall (precipitation – evapotranspiration) received was intercepted by the mole-tile drainage system, with the non-effluent irrigated plot flowing slightly less than the effluent irrigated plot. Table 1 indicates the annual loads and percentage of the annual load that occurred as incidental losses (i.e. occurring during or within a week of an effluent application or grazing event). Results show that drainage water from effluent irrigated plot was more turbid than the non-irrigated plot, but more SS was lost from the non-irrigated plot than the effluent irrigated plot. This indicates that drainage water from the effluent irrigated plot contained SS that was much finer than that from the non-irrigated plot, presumably due to fine suspended material from the effluent. For *E. coli*, while more was lost from the effluent irrigated plot than the non-effluent irrigated plot, the percentage of *E. coli* lost as incidental transfers accounted for the majority of *E. coli* in both plots, except in 2002 in the non-effluent irrigated plot. These incidental transfers were dominated by preferential flow of *E. coli* during 2 events when 2 and 5% of the effluent applied in 2001 was immediately recovered as drainage when application coincided with low soil moisture deficits (Fig. 2). Other studies have shown that nutrient or *E. coli* in drainage waters or overland flow is related to the time since grazing, and that loss of *E. coli* does occur by preferential flow when livestock slurry is applied (e.g., Monaghan et al., 2002; McGechan and Vinten, 2003).

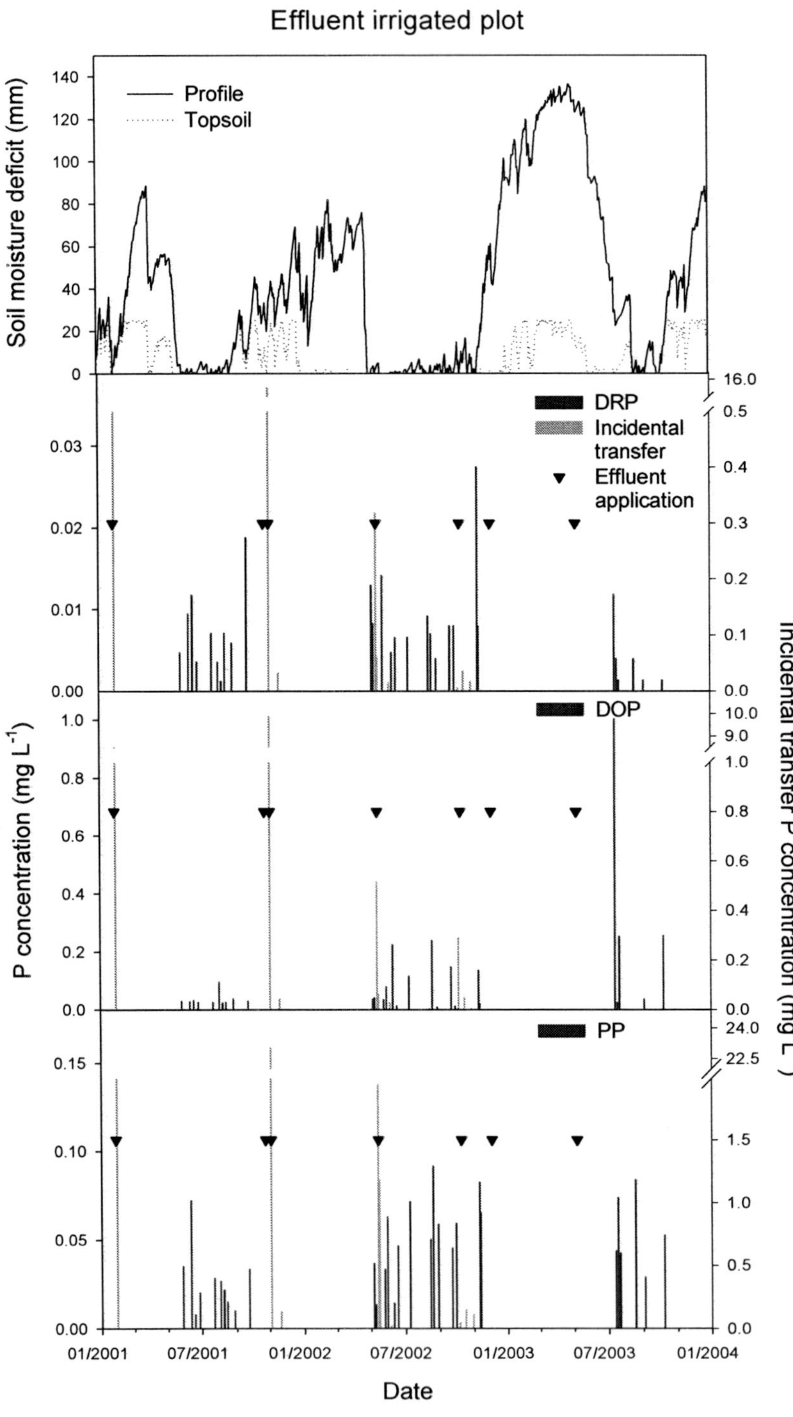

Fig. 2. Soil moisture deficit and concentrations of P fractions in the effluent irrigated plot. Dates of effluent application are indicated by the upside down triangles and events within 7 days of application indicated by the lighter shaded bars

Table 1. Annual loads and the percentage load within 1 week of an effluent application (incidental losses) in effluent and non-effluent plots for nutrients (P fractions, N species), sediments and *E. coli*

Year	Plot	DRP (g ha^{-1})	DOP (g ha^{-1})	TDP (g ha^{-1})	PP (g ha^{-1})	TP (g ha^{-1})	NH_4-N (kg ha^{-1})	NO_3^--N (kg ha^{-1})	*E. coli* (log CFU[†] ha^{-1})	Turbidity (NTU)	TSS[‡] (kg ha^{-1})
Annual											
2001	Non-effluent	13	48	61	102	164	0.03	16.7	10.1	530	5.0
	Effluent	107	122	240	240	469	1.90	21.9	11.6	1177	5.4
2002	Non-effluent	54	87	141	234	375	0.37	45.7	10.3	1091	15.4
	Effluent	54	294	348	239	587	0.53	60.4	11.0	1598	14.1
2003	Non-effluent	2	8	11	49	60	0.04	6.2	8.6	35	0.3
	Effluent	2	121	123	48	171	0.05	8.6	9.2	67	0.5
% Incidental											
2001	Non-effluent	16	9	10	7	8	29	3.5	66.8	9.2	7.9
	Effluent	66	45	55	57	56	68	4.9	97.6	43.5	18.5
2002	Non-effluent	18	15	16	12	13	10	15.0	42.2	26.9	25.8
	Effluent	19	16	16	26	20	10	15.4	69.4	22.5	26.9
2003	Non-effluent	-[§]	-	-	-	-	-	-	-	-	-
	Effluent	-	-	-	-	-	-	-	-	-	-

[†] Coliform forming units.

[‡] Total suspended solids.

[§] No flow events occurred within a week of effluent applications or drainage.

Preferential flow of P during incidental losses accounted for the larger losses of P fractions in the effluent irrigated plot compared to the non-effluent irrigated plot. Indeed, high moisture conditions and the corresponding low soil moisture deficit at this time would have mirrored the saturated conditions under which Jensen et al. (2000) were able to show the preferential flow and loss of P in drainage water from soils treated with cattle faeces. An interesting finding in the study of Jensen et al. (2000) was that during unsaturated conditions most P fractions decreased except DOP, which remained unchanged presumably as many DOP compounds are less strongly sorbed to the soil than orthophosphate (Leytem et al., 2002). Using the lower affinity of DOP for the soil, Jensen et al. (2000) postulated that DOP loss is less dependent upon preferential flow than DRP. Comparing the difference in DOP concentration from each event in our study shows that, excluding DOP loss as a result of incidental transfers, more DOP was lost from the effluent plot than the non-effluent plot as time progressed (Fig. 3). In other words, a significant ($P < 0.05$) relationship was found between the difference in DOP concentration between plots and event number (in chronological order). This corroborates the findings of Jensen et al. (2000) and data in Table 1, which shows an overall increased DOP load from the effluent irrigated plot compared to the non-irrigated plot, but similar losses via incidental transfers in 2002. This suggests that DOP makes up a larger fraction of P losses outside incidental transfers and effluent application appears to have changed the concentration and fractions of P lost, despite both plots receiving similar amounts of P (fertilizer on the non-effluent irrigated plot) to maintain soil Olsen P values. This is an important finding showing that even short-term applications of effluent can alter and promote P loss.

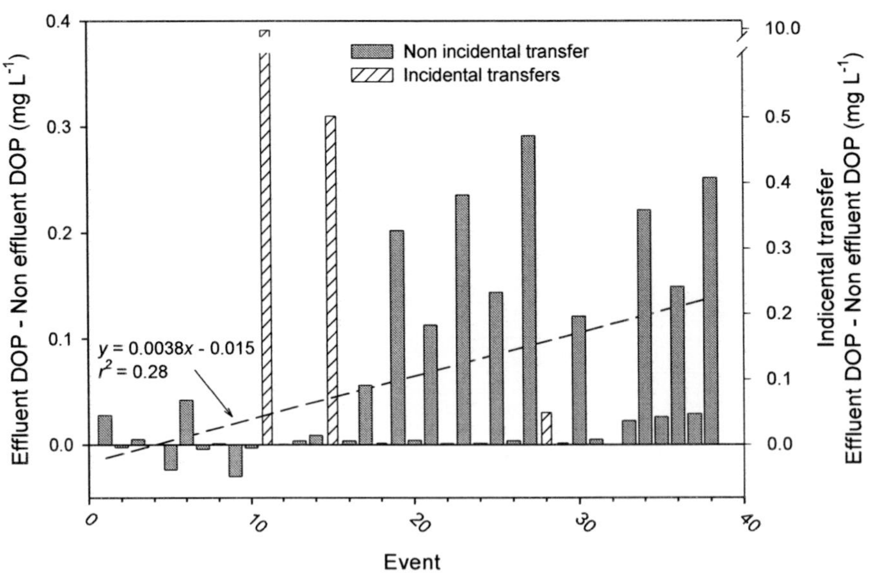

Fig. 3. Relationship between the differences in DOP lost in drainage water from effluent irrigated and non-effluent plots and the sequential number of events (excluding incidental transfers)

Figures 2 and 4 show that the concentrations of DRP, DOP and PP during the three years of the trial were, on average, greater within a week of effluent application or grazing than outside this period (i.e. incidental transfers). Compared to the non-effluent plot, concentrations of DRP in the effluent applied plot exceeded the current New Zealand guidelines for surface water quality more frequently (> 50% of events in excess of 0.01 mg L^{-1}; ANZECC, 2000) in the effluent irrigated plot compared to the non-effluent irrigated plot (*c.* 30% of events in excess of 0.01 mg L^{-1}). While these concentrations are of concern if drainage waters comprise a significant proportion of surface water, several transformations will alter the relative impact of P lost before entering a stream or lake. These include uptake and release by benthic sediments and aquatic biota (e.g., periphyton, microbes), and the sensitivity and trophic status of receiving water bodies (McDowell et al., 2004).

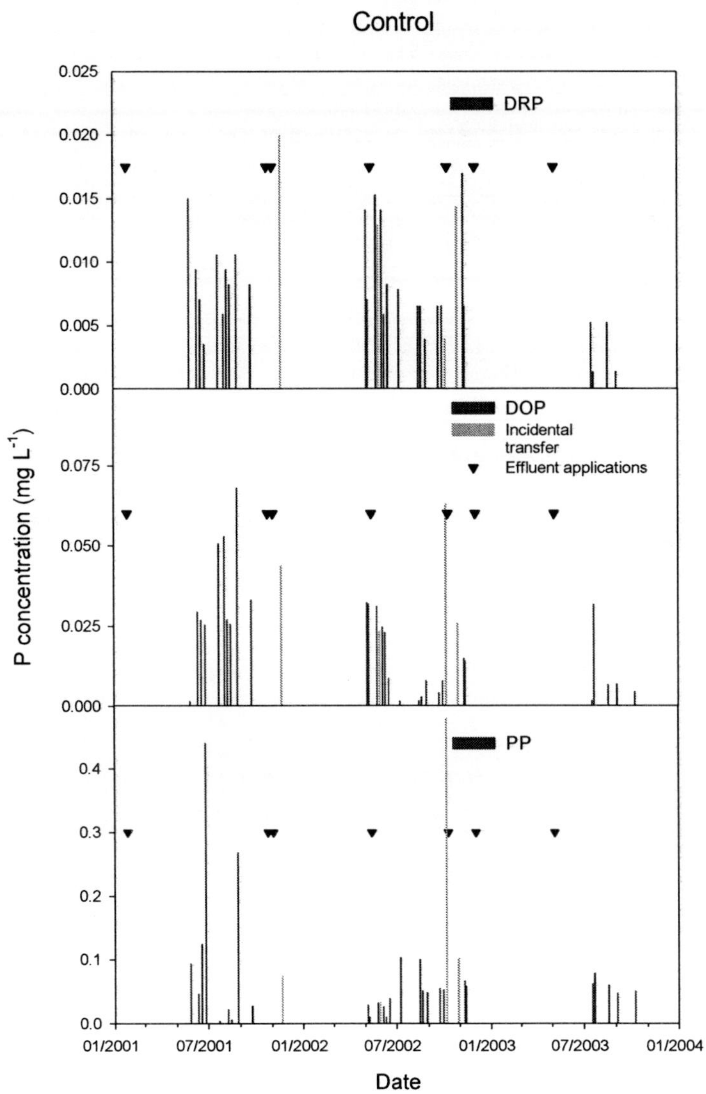

Fig. 4. Concentrations of P fractions in the control (non-effluent) plot. Upside down triangles indicate when effluent was applied to the other plot

Compared to many other studies of P losses from slurry applied to soils, the magnitude of P lost from this system is generally low, except during incidental transfers where DRP concentrations of up to *c.* 16 mg L^{-1} and TP concentrations as high as 40 mg L^{-1} were recorded (Fig. 2). The magnitude of P losses is in line with other studies noting incidental P loss from cattle-grazed grassland systems. For example, Preedy et al. (2001) noted TP losses of 7 and 11 mg L^{-1} from plots that had received a recent application of cattle slurry or cattle slurry + superphosphate. Both examples demonstrate the importance of incidental transfers in cattle-grazed grassland receiving effluent and emphasize the need to schedule effluent applications when soil moisture conditions indicate preferential flow is unlikely.

At this study site it was noted that preferential flow was unlikely when the soil moisture deficit was > 20 mm (Monaghan et al., 2002). Consequently, the potential to mitigate most P loss could simply be achieved by not applying effluent when the soil is this wet. At the other extreme, care should also be taken to ensure that effluent is not applied when the soil is too dry so that large cracks connect the soil surface to drains. Ulén and Mattsson (2003) also concluded that the application of cattle slurry should only occur during dry conditions. However, this does not impact on the effect of effluent application on the bulk of the soil and also the fact that more applications, even in the short-term, caused more P to be lost compared with the non-irrigated plot, even though soil Olsen P concentrations were similar (Tables 1 and 2).

Several reasons may account for enhanced P loss from effluent irrigated soils. These are primarily due to either a build-up of labile and mobile organic P forms or decreased sorption capacity of bulk soil and/or preferential flow pathways due to competition by organic compounds or blockage by P-rich colloids (McDowell et al., 2004). A number of techniques were employed to test the chemical P fractions present and the response of effluent and non effluent irrigated soils to further P additions.

Soil Analyses

Data for soil chemical characteristics, P fractions, and organic P characterization via liquid state ^{31}P NMR in the effluent irrigated and non-effluent irrigated plot for the 0-2, 0-7.5, 7.5-15 and 15-45 cm depths are given in Tables 2, 3 and 4. Although Olsen P was similar in effluent and non-effluent plots, pH and total base saturation (TBS) was higher in the top two depth increments of the effluent plot than the non-effluent plot (Table 2). It is well known that an increase in soil pH is reflected by TBS, and that the retention of P, and metal-humic acid complexes such as those that contain organic P, is pH dependant (e.g., Tan, 1998). The increase in soil pH is probably due to the application of large amounts of Ca and Mg in effluent, and since the amount of P held by metal-humate complexes decreases with increasing pH, this may account for some enhancement of P loss in the effluent plot.

Table 2. Mean soil chemical properties for non-effluent and effluent treated soils at different depths

Depth (cm)	Exchangeable bases				Olsen P	TBS[†] (%)	Organic C	Organic matter	pH	Freundlich Sorption parameters	
	Ca ($cmol_c\ kg^{-1}$)	K ($cmol_c\ kg^{-1}$)	Mg ($cmol_c\ kg^{-1}$)	Na ($cmol_c\ kg^{-1}$)	($mg\ kg^{-1}$)	(%)	($g\ kg^{-1}$)	($g\ kg^{-1}$)		α	β
Non-effluent treated soils											
0-2	12	10	26	11	124	59	49	86	5.8	16.4	1.018
0-7.5	12	9	20	8	84	57	44	76	5.7	37.9	1.316
7.5-15	10	4	10	5	25	56	33	59	5.7	97.7	1.918
15-45	6	3	7	5	16	42	15	31	5.7	48.2	2.321
Effluent treated soils											
0-2	13	23	27	11	122	66	50	85	5.9	16.1	1.082
0-7.5	12	15	30	15	89	64	47	82	5.8	27.2	1.224
7.5-15	10	5	15	5	31	54	34	57	5.7	88.2	1.742
15-45	6	4	7	4	16	43	18	26	5.7	33.4	2.168
LSD[‡]	1	2	2	1	4	2	2	2	0.1	-[§]	-

[†] Total base saturation.

[‡] At the $P < 0.05$ level of significance

[§] LSD not calculated as only duplicates used to generate sorption isotherms

Table 3. Chemical fractionation of P forms (mg kg^{-1}) in non-effluent and effluent treated soils at different soil depths

Depth (cm)	H_2O	H_2O	Ca-EDTA	Ca-EDTA	Na-EDTA	Na-EDTA	ASOP	Cold TCA	Hot TCA	NaOH	NaOH	Residual	Total
	Pi[†]	Po[‡]	Pi	Po	Pi	Po				Pi	Po		
Non-effluent treated soils													
0-2	32.7	9.8	221	31	389	98	77	9.6	20.4	93	174	4.4	1142
0-7.5	20.6	7.9	161	41	300	102	89	12.5	12.2	91	177	2.4	997
7.5-15	2.6	5.2	74	64	137	72	52	6.3	9.0	76	153	1.6	657
15-45	0.8	1.2	54	63	82	40	55	4.6	6.8	49	121	4.2	481
Effluent treated soils													
0-2	42.2	12.6	243	24	430	81	74	7.8	22.0	96	201	3.1	1249
0-7.5	34.8	8.2	180	32	349	82	70	8.3	21.4	90	206	3.2	1086
7.5-15	4.1	4.8	102	25	140	89	60	6.2	14.2	76	159	3.3	682
15-45	0.5	1.4	56	14	77	63	50	4.8	8.7	49	132	3.2	459
LSD[§]	3.2	2.0	19	31	35	21	22	3.4	3.0	7	13	1.3	37

[†] Inorganic P

[‡] Organic P

[§] At the $P < 0.05$ level of significance

However, an increase of 0.1 pH units, while significant, is probably not great enough to singularly account for the larger losses of DOP from the effluent plot compared to the non-effluent plot. Data for the sequential P fractionation indicated that a greater concentration of water soluble P was in organic P form in topsoil (0-7.5 cm) in effluent irrigated than non-irrigated soils (Table 3). In a ^{31}P NMR analysis of leachate from soil columns treated with dairy effluent, Toor et al. (2003) found that most P was in the form of monoesters and diesters (> 80% of total P). The collective studies of Leytem et al. (2002), Frossard et al. (1989), and Hoffman and Rolston (1980) have shown that organic P compounds, especially diesters, are generally more labile and mobile than orthophosphate. The exception was phytic acid, a monoester, which was shown to be more strongly sorbed to soil than orthophosphate. Given that the soil used in the study of Toor et al. (2003) was of the same soil group, a Hapludalf (a Pallic soil in NZ soil classification), it is possible that DOP in this study is being lost as these mobile organic P forms. However, other soil P analyses, soil P fractionation data and the analysis of NaOH-EDTA extracts by ^{31}P NMR do not support this conclusion as the sole reason for enhanced DOP loss in the effluent irrigated plot. Other mechanisms are also likely involved.

Soil P fractionation data show that more P is accumulated in the effluent irrigated soil down to 7.5 cm depth than in the non-effluent irrigated soil (Table 3). Most of this accumulation occurred in the Ca-P fraction (Na-EDTA), probably due to the application of large amounts of Ca in the effluent. In contrast to Siddique and Robinson (2003), who examined P availability resulting from animal manure applications to a soil of near neutral pH (7.1 in water), no decrease in H_2O-P was noted with increasing Ca concentrations in our effluent irrigated soil.

Organic P fractions obtained by the sequential extraction procedure (Fe-P [Na-EDTA], Ca-P, ASOP, cold and hot TCA and NaOH) are similar between effluent and non-effluent irrigated soils. The exception is for the NaOH fraction. This fraction represents largely humic material and contains much phytate (De Groot and Golterman, 1993). A comparison of ^{31}P NMR data for effluent and non-effluent soils shows that more monoesters, of which phytate makes up a sizeable proportion, are found in the effluent irrigated soils (Table 4). Indeed, a significant relationship exists between NaOH-Po and monoester concentration (monoesters = 2.2 × NaOH Po − 18, $R^2 = 0.66$, $P < 0.05$). A typical example of a ^{31}P NMR spectrum is shown in figure 5, showing the large proportion of P in the monoester range. Data from an analysis of fresh effluent also shows that while most P is as orthophosphate, organic P compounds were as monoesters, most probably phytate (Fig. 1). Recent evidence has shown the presence and dominance of phytate in the monoester region of ^{31}P NMR spectra of NaOH-EDTA extracts of manures (Turner, 2004), and since phytate is more strongly sorbed than orthophosphate (Leytem et al., 2002), phytate may out-compete and deter further P sorption.

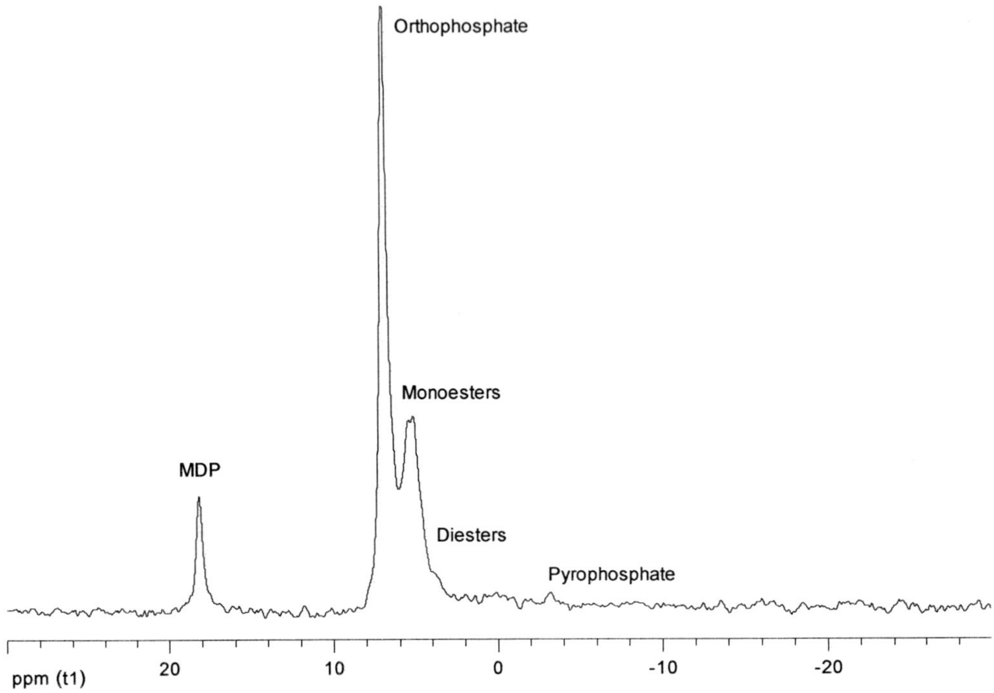

Fig. 5. ^{31}P NMR spectra of a NaOH-EDTA extract of the 0-2cm depth of the effluent applied plot. MDP is the reference standard

Effluent contains much material other than P. Analysis of the effluent by LECO© C/N analyzer indicated that most (> 80 g kg^{-1}) was present as organic matter. The remainder is most likely comprised of the inorganic component of small colloids. Both organic matter and colloids will travel through preferential flow pathways and into the soil matrix itself (Geohring et al., 2001). Addition of 1 mg P L^{-1} to the > 0.45 μm fraction (overnight shaking, centrifugation at 4000 × *g*) of effluent indicated that you got more than 1 mg P L^{-1} released back into solution. The colloidal material is P saturated and will likely deter additional P sorption in flow pathways, should it lodge there. Similarly, since much organic matter is negatively charged, many organic compounds can compete with P for sorption sites. Ohno and Crannell (1996) showed that dissolved organic matter (DOM) from common green manures (e.g., *Trifolium incarnatum* sp.) and organic acids (e.g., citric acid) can compete and deter P sorption via ligand exchange reactions with soil Al. Similarly, Fox and Comerford (1990) found that low-molecular-weight organic acids from the soil solution of forested Utlisols, Entisols and Spodosols (e.g., Oxalic and citric acid) dislodged P from sorption sites thereby making P available to the plant. However, Ohno and Crannell (1996) also showed that DOM from cattle and poultry manure had little effect on P sorption. Afif et al. (1995) showed that organic matter delayed but did not prevent P sorption in 12 soils (mainly Oxisols). Studies that have removed organic matter from the soil have shown increased P sorption afterwards, but may have also altered the inorganic component of the soil (e.g., H_2O_2 treatment; Bhatti et al., 1998). However, McDowell and Condron (2001) found that removing organic matter with NaOCl from manured and non-manured grassland soils increased P sorption in the unmanured soil relative to the manured soil, and attributed this to more resilient organic matter in manured soil and the removal of organic matter not active in P sorption. Further investigation is warranted since the effect of DOM on P sorption and desorption is unclear.

Table 4. Concentration (mg kg^{-1}), and percentage in parentheses, of P forms in non-effluent and effluent treated soils at different soil depths as detected by ^{31}P liquid state NMR

Depth (cm)	Ortho-phosphate	Monoesters	Diesters	Pyro-phosphate	Poly-phosphate	Phosphonates
	(6.26)†	(3 to 6)‡	(1 to -1)	(-3 to -6)	(-20)	(20)
Non-effluent treated soils						
0-2	562 (54)	386 (37)	63 (6)	25 (2)	-	-
0-7.5	440 (52)	367 (45)	36 (4)	-	-	-
7.5-15	241 (42)	325 (57)	5 (1)	-	-	-
15-45	151 (36)	270 (64)	-	-	-	-
Mean	349	337	35	-	-	-
SEM§	93	26	15	-	-	-
Effluent sample						
	16.7 (88)	2.4 (13)	-	0.1 (0.4)	-	-
Effluent treated soils						
0-2	557 (52)	450 (42)	44 (4)	11 (1)	-	-
0-7.5	489 (56)	351 (41)	16 (2)	6 (1)	-	-
7.5-15	291 (49)	294 (49)	4 (1)	4 (1)	-	6 (1)
15-45	176 (43)	217 (54)	-	3 (1)	6 (2)	-
Mean	378	328	21	6	-	-
SEM	88	49	10	2	-	-

† Chemical shift (δ ppm).
‡ Total percentage >100 due to rounding.
§ Standard error of the mean.

Overall, in the effluent irrigated plot, data for the fit of the Freundlich equation to P sorption isotherms indicated that both parameters α and β were lower than their non-effluent applied counterparts. This indicates that the sorption capacity and strength of the effluent irrigated soils has decreased compared to the non-effluent plot. While the decrease is not great, Holford et al. (1997) noted that P leaching down the soil profile would occur well before the total sorption capacity had been saturated by P, largely due to a lowering of sorption strength. The lower sorption strength of effluent irrigated soils will have made the lesser sorption of DOP compounds in effluent irrigated soil more pronounced than in the non-effluent soil.

Sorption-Desorption Reactions

A comparison of effluent irrigated and non-effluent soils at each depth and their response to a number of P additions is given in Figure 6. All data has had data for the control soil subtracted before analysis. The data shows that, as expected, more P is sorbed by deeper soil layers. This applies to both effluent irrigated and non-effluent soils. However, less P is generally sorbed in the effluent irrigated soils reflecting a higher P concentration and greater occupancy rate of P sorption sites than in the non-effluent soils (Fig. 6). In terms of desorption, more DRP and DOP was generally released from the effluent irrigated soils than the non-effluent soils. However, it is worth noting that DOP desorption was much more pronounced in the 0-2 cm depth than in other depths of non-effluent soils. This is where the greatest P, organic C and organic matter concentrations in the soil profile are located and will have impacted on DOP release due to a subsequent lesser sorption strength and capacity (Tables 2 and 3).

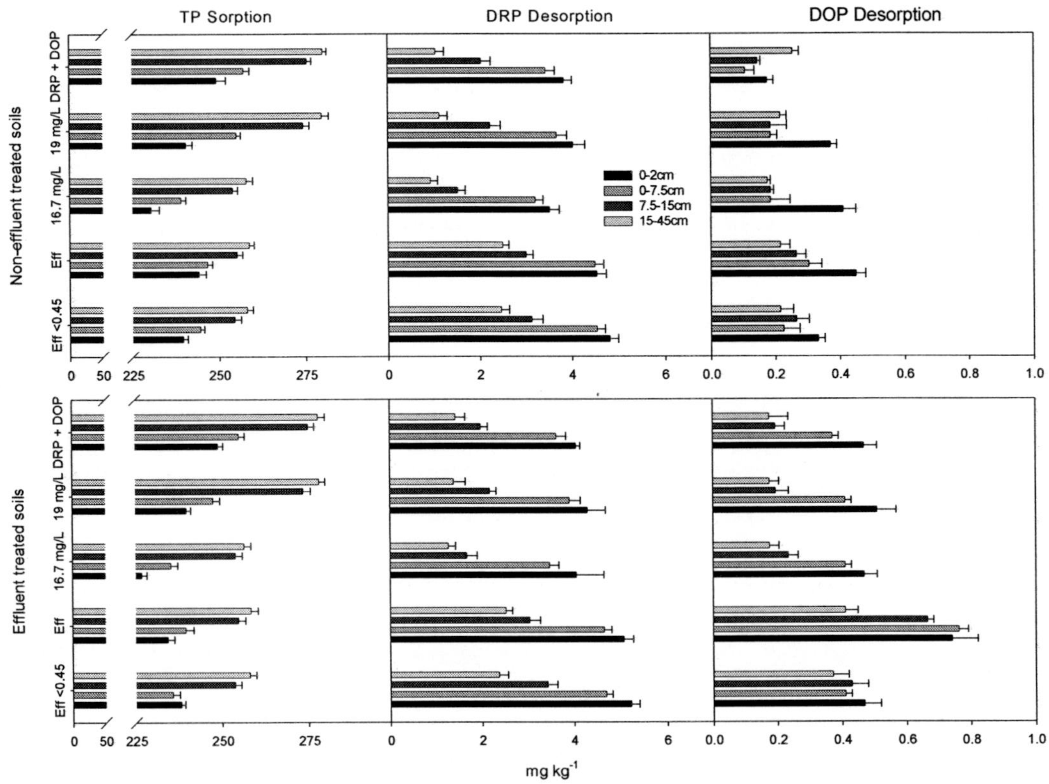

Fig. 6. Total P (TP) sorbed followed by the desorption of DRP and DOP in effluent and non-effluent treated soils subject to a range of treatments (orthophosphate at effluent DRP concentration [16.7 mg L^{-1}], orthophosphate at effluent TDP concentration [19 mg L^{-1}], orthophosphate + known organic P compounds at effluent DRP + DOP concentration [DRP + DOP], effluent < 0.45 μm [Eff < 0.45], and whole fresh effluent [Eff]). Error bars represent 95% confidence intervals

Comparing P treatments in the non-effluent soils, more P was sorbed in those supplied with more P as orthophosphate, the 19 mg P L^{-1} treatment, compared to the 16.7 mg P L^{-1} treatment (Fig. 6). Similar concentrations were sorbed by the 7.5-15 and 15-45 cm depths for the two effluent treatments in non-effluent and effluent irrigated soils. However, in these treatments more P was sorbed in the top two depth increments suggesting increased organic matter and binding via ligands may have assisted. Since the data is presented as TP sorbed, the enhanced sorption of phytate over orthophosphate may also have assisted P sorption. Certainly, this would explain why the synthetic effluent treatment (DRP + DOP), which contained a significant proportion of P as phytate exhibited the greatest P sorption of all treatments.

For the non-effluent irrigated soils, desorption of DRP mirrored P sorption in that more DRP was desorbed from the treatment receiving 19 mg P L^{-1} than the treatment receiving 16.7 mg P L^{-1}, both as orthophosphate. Interestingly, the DRP desorbed from the synthetic effluent treatment was similar to the 19 mg P L^{-1} treatment even though 2.3 mg P L^{-1} of total P (12%) in the synthetic effluent treatment was applied as phytate, pyrophosphate and polyphosphate. This supports the enhanced sorption of phytate and displacement of orthophosphate (Leytem et al., 2002), especially since the concentration of DOP in the DRP + DOP treatment was less than that from the 19 mg P L^{-1} orthophosphate treatment. Furthermore, more DRP is desorbed from the two effluent treated soils compared to the other treatments, even in effluent irrigated soils. This suggests that something other than exclusively P compounds are influencing P desorption in our soils. The similarity in DRP desorption data from the two effluent treated soils also shows that this is unlikely to be related to the > 0.45 μm fraction.

Desorption of DOP in non-effluent irrigated soils was similar among all treatments in the 0-7.5, 7.5-15, and 15-45 cm depths. However, differences were noted between these depths and the 0-2 cm depth. Desorption of DOP was also more pronounced in the effluent treated soils compared to other treatments in both non-effluent and effluent irrigated soils. This suggests two things: firstly, that a component of the > 0.45 μm fraction is influencing DOP desorption, and secondly, that this is further aided by a function concentrated in the 0-2 cm of topsoil. In non-effluent irrigated soils, the only effluent-equivalent is dung via grazing cattle. Since this is in a very concentrated form and unlikely to seep below the topsoil, we argue that retention of dung in the top 0-2 cm is influencing DOP desorption in the non-effluent irrigated topsoil (0-2 cm). Comparative organic C and organic matter concentrations also feature in the 0-2 cm depths of effluent irrigated and non-effluent soils, but not at other depths (Table 2). It is also possible that P from saturated colloids may have re-equilibrated with P in soil and influenced P desorption. Maguire et al. (2002) showed that small particles are enriched with labile P relative to large particles, and that the suspension of different aggregate sizes in solution causes a re-distribution of water soluble P from P-rich aggregates to a P-poor aggregates. Since many DOP compounds are less strongly held to the soil than orthophosphate, *any* effect on desorption will influence DOP more than DRP.

Digesting all this data highlights a number of findings:

1. Highly P sorptive P compounds such as phytate influence the sorption and desorption of DRP, but the effect on DOP desorption is unclear.
2. Compounds in effluent/dung other than P influence DRP and DOP desorption.
3. Compounds in particles > 0.45 μm influence DOP desorption.

While the first two of these findings have mechanisms that have been documented in the literature and explained earlier in this study, the third finding, the influence of colloidal compounds on P desorption warrants further attention. At this time we can only speculate that adding colloids saturated with P and which contain many organic compounds either dislodge or out compete P, especially DOP, from P sorption sites. Both mechanisms facilitate DOP desorption by influencing the soil matrix and flow pathways and displacing DOP that is less strongly held by soil than orthophosphate.

CONCLUSIONS

The short-term application of dairy effluent to a mole-pipe drained grazed pasture caused more P to be lost, especially during incidental transfers, than a plot of equivalent Olsen P concentration, but receiving P as superphosphate. Data suggest much of the P lost from effluent irrigations could be mitigated by better scheduling applications when the soil can soak the liquid up without risking preferential flow. However, analysis of P forms indicates that more P is being lost, especially as DOP, from the effluent irrigated plot via non-incidental transfers as the number of applications and events increase. Analysis of both effluent irrigated and non-irrigated plot soils indicated more P was held in the Ca-P fraction, water soluble P fraction and NaOH organic P fraction. This was due to the application of a large amount of Ca in the effluent, which also significantly increased soil pH. The increase in organic P in the NaOH fraction was attributed to the application of phytate via effluent. Since phytate is more strongly sorbed to soil than orthophosphate and P sorption strength decreases with increasing pH, both mechanisms were attributed to the overall decrease in P sorption capacity and strength and subsequent increased P losses. A third mechanism, the re-distribution of P from colloids and competition of P sorption sites by organic compounds was hypothesized as further enhancing P loss and warrants attention. Since DOP compounds are less strongly held than DRP (orthophosphate), we conclude any mechanisms influencing P retention by the soil matrix or in soil lining preferential flow pathways will compound to promote DOP loss more than DRP loss. This highlights the need to quantify DOP losses and bioavailability, especially in effluent treated soils.

ACKNOWLEDGEMENTS

Funding for this work was provided by the New Zealand Foundation for Research, Science and Technology under contract AGRX002.

REFERENCES

ANZECC. 2000. Australian and New Zealand Guidelines for Fresh and Marine Water Quality. Australian and New Zealand Environment and Conservation Council – Agriculture and Resource Management *Council of Australia and New Zealand.* Canberra, ACT.

Afif, E., V. Barrón, J. Torrent. 1995. Organic matter delays but does not prevent phosphate sorption by Cerrado soils from Brazil. *Soil Sci.* 159:207-211.

Bhatti, J.S., N.B. Comerford. And C.T. Johnston. 1998. Influence of oxalate and soil organic matter on sorption and desorption of phosphate onto a Spodic horizon. *Soil Sci. Soc. Am. J.* 62:1089-1095.

Cade-Menun, B.J., and C.M. Preston. 1996. A comparison of soil extraction procedures for ^{31}P NMR spectroscopy. *Soil Sci.* 161:770-785.

Chardon, W.J., O. Oenema, P. del Castilho, R. Vriesema, J. Japenga, and D. Blaauw. 1997. Organic phosphorus in solutions and leachates from soils treated with animal slurries. *J. Environ. Qual.* 26:372-378.

De Groot C.J., and H.L. Golterman. 1993. On the presence of organic phosphate in some Camargue sediments: evidence for the importance of phytate. *Hydrobiologia* 252:117-126.

Eghball, B., G.D. Binford, and D.D. Baltensperger. 1996. Phosphorus movement and adsorption in a soil receiving long-term manure and fertilizer application. *J. Environ. Qual.* 25:1339-1343.

Fox, T.R., and N.B. Comerford. 1990. Low-molecular-weight organic acids in selected forest soils of the Southeastern USA. *Soil Sci. Soc. Am. J.* 54:1139-1144.

Frossard, E., J.W.B. Stewart, R.J. St Arnaud. 1989. Distribution and mobility of phosphorus in grassland and forest soils of Saskatchewan. *Can. J. Soil Sci.* 69:401-416.

Geohring, L.D., O.V. McHugh, M.T. Walter, T.S. Steenhuis, M.S. Akhtar, and M.F. Walter. 2001. Phosphorus transport into subsurface drains by macropores after manure applications: implications for best manure management practices. *Soil Sci.* 166:896-909.

Golterman, H.L. 1960. Studies on the cycle of elements in fresh water. *Acta Bot.* 9:1-58.

Golterman, H.L. 1996. Fractionation of sediment phosphate with chelating compounds: a simplification, and comparison with other methods. *Hydrobiologia* 335:87-95.

Golterman, H., J. Paing, L .Serrano, and E. Gomez. 1998. Presence of and phosphate release from polyphosphate or phytate phosphate in lake sediments. *Hydrobiologia* 364:99-104.

Gómez, J.C.C., and F.J.S. López. 2004. Mestre-C, Nuclear Magnetic Resonance Companion. v. 3.5.1. Available at http://www.mestrec.com (verified Feb. 2004).

Hoffman, D.L., and D.E. Rolston. 1980. Transport of organic phosphate in soil as affected by soil type. *Soil Sci. Soc. Am. J.* 44:46-52.

Holford, I.C.R., C. Hird, and R. Lawrie. 1997. Effects of animal effluents on the phosphorus sorption characteristics of soils. *Aust. J. Soil Res.* 35:365-373.

Jensen, M.B., T.B. Olsen, H.C.B. Hansen, and J. Magid. 2000. Dissolved and particulate phosphorus in leachate from structured soil amended with fresh cattle faeces. *Nut. Cyc. Agroecosyst.* 56:253-261.

Leytem, A.B., Mikkelsen, R.L. and J.W. Gilliam. 2002. Sorption of organic phosphorus compounds in Atlantic coastal plain soils. *Soil Sci.* 167:652-658.

Maguire, R.O., A.C. Edwards, J.T. Sims, P.J.A. Kleinman, and A.N. Sharpley. 2002. Effect of mixing soil aggregates on the phosphorus concentration in surface waters. *J. Environ. Qual.* 31:1294-1299.

McDowell, R.W., and L.M. Condron. 2001. Influence of soil constituents on soil phosphorus sorption and desorption. *Commun. Soil Sci. Plant Anal.* 32:2531-2547.

McDowell, R.W., and R.M. Monaghan. 2002. The potential for phosphorus loss in relation to nitrogen fertilizer application and cultivation. *N.Z. J. Agric. Res.* 45:245-253.

McDowell, R.W., B.J.F. Biggs, A.N. Sharpley, L. Nguyen. 2004. Connecting phosphorus loss from agricultural landscapes to surface water quality. *Chem. Ecol.* 20:1-40.

McGechan, M.B., and J.A. Vinten. 2003. Simulation of transport through soil of *E. coli* derived from livestock slurry using the MACRO model. *Soil Use Manage.* 19:321-330.

MfE. 2002. Microbiological Water Quality Guidelines for Marine and Freshwater Recreational Areas. *Ministry for the Environment*, New Zealand Government, Wellington, New Zealand.

Monaghan, R.M., Horne, D, M. Hedley, and L. Nguyen. 2002. Subsurface drainage review: water and contaminant loads from artificial subsurface drainage systems in New Zealand. *Report for Fonterra Research Centre, AgResearch Invermay*, Mosgiel, New Zealand.

Ohnno, T., and B.S. Crannell. 1996. Green and animal manure-derived dissolved organic matter effects on phosphorus sorption. *J. Environ. Qual.* 25:1137-1143.

Olsen, S.R., C.V. Cole., F.S. Watanabe, and L.A. Dean. 1954. Estimation of available phosphorus in soils by extraction with sodium bicarbonate. United States Department of Agriculture Circular No 939. *United States Government Print Office*, Washington, USA.

Preedy, N., K. McTiernan, R. Matthews, L. Heathwaite, and P. Haygarth. 2001. Rapid incidental phosphorus transfers from grassland. *J. Environ. Qual.* 30:2105-2112.

Ruttenberg, K.C. 1992. Development of a sequential extraction method for different forms of phosphorus in marine sediments. *Limnol. Oceanogr.* 37:1460-1482.

Siddique, M.T., and J.S. Robinson. 2003. Phosphorus sorption and availability in soils amended with animal manures and sewage sludge. *J. Environ. Qual.* 32:1114-1121.

SPSS Inc. 1999. *SPSS,* Version 10.0. Chicago, IL.

Sumner, M.E., and W.P. Miller. 1996. Cation exchange capacity and exchange coefficients. p. 1201-1229. *In* D.L. Sparks, A.L. Page, P.A. Helmke, R.H. Loeppert, P.N. Soltanpour, M.A. Tabatabai, C.T. Johnston, and M.E. Sumner (ed.) Methods of Soil Analysis, Part 3, chemical methods, SSSA book series no. 5. *Soil Science society of America*, American Society of Agronomy, Madison, WI.

Tan, K.H. 1998. Principles of Soil Chemistry. *3rd Ed. Marcel Dekker, Inc.* New York, NY.

Toor, G.S., L.M. Condron, H.J. Di, K.C. Cameron, B.J. Cade-Menun. 2003. Characterization of organic phosphorus in leachate from a grassland soil. *Soil Biol. Biochem.* 35:1317-1323.

Turner, B.L. 2004. Optimizing phosphorus characterization in animal manures by solution phosphorus-31 nuclear magnetic resonance spectroscopy. *Soil Sci. Soc. Am. J.* 68: In press.

Turner, B.L., I.D. McKelvie, P.M. Haygarth. 2002. Characterisation of water-extractable soil organic phosphorus by phosphatase hydrolysis. *Soil Biol. Biochem.* 34:27-35.

Turner, B.L., N. Mahieu, L.M. Condron. 2003. Phosphorus-31 nuclear magnetic resonance spectral assignments of phosphorus compounds in soil NaOH-EDTA extracts. *Soil Sci. Soc. Am. J.* 67:497-510.

Ulén, B. and L. Mattsson. 2003. Transport of phosphorus forms and of nitrate through a clay soil under grass and cereal production. *Nut. Cyc. Agroecosyst.* 65:129-140.

Watanabe, F.S., and S.R. Olsen. 1965. Test of an ascorbic acid method for determining phosphorus in water and $NaHCO_3$ extracts from soil. *Soil Sci. Soc. Am. Proc.* 29:677-678.

Whitton, B.A., S.L.J. Grainger, G.R.W. Hawley, and J.W. Simon. 1991. Cell-bound and extracellular phosphatase activities of cyanobacterial isolates. *Microbiol. Ecol.* 21:85-98.

In: Water Pollution: New Research
Editor: A. R. Burk, pp. 77-98

ISBN: 1-59454-393-3

Chapter 4

EFFECTS OF SAMPLING AND SAMPLE PREPARATION IN BIOINDICATION OF INVERTEBRATES

***Bettina Zödl,*[*,1] *Michaela Zeiner*[2] *and Claudia Gundacker*[3]**

[1]Institute of Physiology, Dptm. Environmental Physiology and Balneology, Medical University of Vienna, Schwarzspanierstraße 17, A-1090 Vienna, Austria
[2]Institute of Analytical Chemistry, University of Vienna, Währingerstrasse 38, A-1090 Vienna, Austria
[3]Institute of Medical Biology, WG Ecophysiology and Ecotoxicology, Medical University of Vienna, Währingerstraße 10, A-1090 Vienna, Austria

ABSTRACT

In recent years invertebrates are gaining importance for bioindication due to their integration of pollutants in space and time, the short life cycle and the relative ease of collection. Nevertheless, it is necessary to reflect upon the methodology of collecting and sample preparation as previous metal studies on invertebrates, living in urban environments, suggest that a variety of factors (e.g. trophic level, strategies for detoxification, defecation, and type of trapping) are responsible for differences in sample concentrations. In order to obtain basic information for designing standardised test preparation methods, the heavy metals Zn, Cu, Cd and Pb were measured in gastropods (*Xerolenta obvia*), oligochaetes (*Lumbricus terrestris*), isopods (*Armadillidium vulgare* and *Trachelipus rathkei*) and carabids (*Harpalus rubripes* and *Calathus fuscipes*) using different sampling methods and different modes of sample treatment. Isopods and carabids trapped with formalin-pitfalls were compared to manually collected specimens (which were allowed to defecate). Additionally effects of defecation were studied by feeding cellulose as an accelerator of gut passage in gastropods and oligochaetes. Effects of sample cleaning were determined in isopods (*A. vulgare*) between deionate-washed and ultrasonic-cleaned individuals. Type of sample drying is very important as volatile

* Corresponding author: Bettina Zödl: Medical University of Vienna, Institute of Physiology. Department of Environmental Physiology and Balneology, Schwarzspanierstrasse 17. A-1090 Vienna. Austria. Tel.: +43-1-4277-62114. Fax: +43-1-4277-62199. e-mail: bettina.zoedl@meduniwien.ac.at

elements like Cd, Hg, and Se may exhaust during sample preparation. For that reason Cd concentration was compared between heat-dried versus freeze-dried isopods. Sex-specific differences in metal concentrations were investigated in carabids. Based on our and previous results, invertebrates should be: collected *in vivo*, allowed to defecate, be freeze-fixed and (at least arthropods) ultrasonic-cleaned, determined to species level and in certain groups (carabids) also to sex, and then be sized or sorted by size (and age) before further preparation and analysis. If any of these treatments is impractical, comparable sampling and preparation methods are recommended as a minimum requirement in order to avoid bias in the results and/or interpretation.

Key words: cadmium, lead, copper, zinc, AAS, defecation, pitfall, cleaning, drying

THEORETICAL BACKGROUND

Biomonitoring

Bioindicators are organisms or communities of organisms which react to environmental effects (e.g. pollutant load) by changing their vital functions and/ or their chemical composition (e.g. by accumulating pollutants) thus making it possible to draw conclusions on the state of their environment [1].

The most important reasons for using bioindicators in monitoring environmental pollution are [2,3]:

- The direct determination of biological effects: Under field conditions lower concentrations of pollutants cause a biological effect than under standardised laboratory conditions
- Bioavailability as well as chronic exposure to pollutants is considered
- The determination of effects of multiple pollutants on organisms
- Metabolism of pollutants by organisms can be considered
- Content of the pollulant can be integrated by time in the organisms' bodies
- Effects of food chain like biomagnification are visible
- Indirect consequences by complicated ecological interactions (like predator-prey) can be considered
- Relatively low costs compared to technical/chemical measuring methods

In principle bioindicators are grouped into accumulation indicators and response indicators. Accumulation indicators store pollutants, which can be measured by chemical methods. Most of the topic of this chapter will deal with that part.

Respond indicators react with changes from molecular biological up to ecological level when taking up even small amounts of harmful substances. Changes in different organisational levels can be enzymatic or membrane functions at the biochemical cellular level, changes in the thickness of eggshells up to changes in chronobiology like fall of leaves or activity phases. [cf. 1-3]

Response and accumulation indicators are subdived into three different groups:

- Pointer organisms give evidence on the state of entire ecosystems
- Test organisms, which are used under highly standardised laboratory conditions
- Monitoring organisms are used to monitor both quality and quantity of harmful substances in the environment.

There is a principial distinction between active and passive biomonitoring. Passive bioindicators are a natural already existing component of the ecosystem, whereas active biomonitoring includes all methods which insert organisms under standardised conditions into the site to be monitored of the ecosystem.

In this chapter we will describe aspects of sampling and sample preparation of passive accumulating bioindicators (invertebrates). [cf. 1]

There are various criteria for bioindicators which have to be fulfilled:

- Wide spread and frequently existing
- Easy sampling and identification
- Small habitat (marginal cruising radius)
- Possibility of standardisation of sampling, sample preparation and measurement
- Adequate size for routine examination
- Equal food over a longer period
- Medium sensitiveness
- Correlation between concentration of the pollutant in the food and body
- Species should be suitable for additional laboratory studies
- Enough biological/ecological information about the species [cf. 2,4]

Invertebrates as Bioindicators

In pollution studies with organisms, living in urban environments, invertebrates have more often been used as passive, indirect bioindicators or accumulation indicators rather than as active, direct ones. Advantages of using invertebrates for bioindication studies are the integration of pollutants in space, the short life cycle and the relative ease of collection. Some invertebrate species like isopods, earthworms and gastropods (especially aquatic ones) accumulate metals to high levels (e.g. isopods strongly accumulate Cu), and therefore are used in applied issues as bioindicators in assessment of environmental metal pollution [5-8]. Nevertheless, it is necessary to reflect on the methodology of collecting and sample preparation and to examine the sensitivity (and the habitat) of each species before using it for bioindication. Previous metal studies on invertebrates at urban sites revealed that a variety of factors are responsible for differences in sample concentrations. Concentrations of metals may increase with higher trophic level [9,10] or show positive (or as well negative) correlations with animal weight [11] or age [12,13]. In molluscs metals are accumulated to higher concentrations in soft bodies than in shells [14]. Regarding age successive developmental stages showed different Cd and Pb concentrations in whole individuals of *Lymantria dispar* [15].

Differences in metal concentrations of invertebrates may also reflect different strategies for detoxification, e.g. storage, moulting, excretion or binding to metallothioneins [9,11,16-18]. Highest concentrations of Pb, Cu and Zn were found in the byssal threads of *Dreissena polymorpha,* which might be an effective strategy for preventing toxic injury [19]. As the size (or weight or age) of an organism may influence the bioaccumulation of pollutants the author size-standardised the different metal concentrations in mussel bodies considering the shell length [19].

Sample Preparation

Some common steps involved in the process of chemical analysis are shown in figure 1.

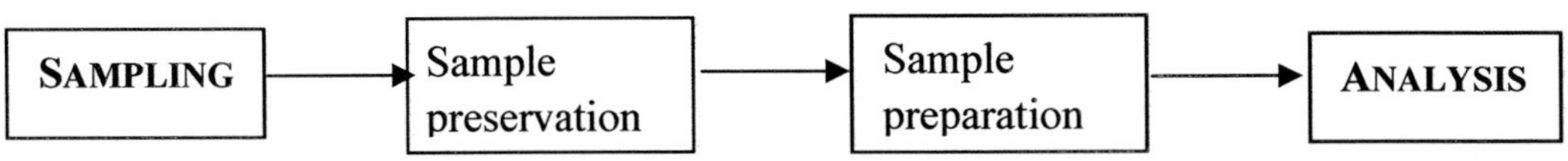

Figure 1: Steps in a measurement process

An error in the first three steps cannot be rectified by even the most sophisticated analytical instrument. So, the importance of the prior steps, in particular the sample preparation, should not be underestimated. Prior to sampling, sample size should be calculated, considering the high standard deviations of biological samples, in oder to prevent bias of statistical interpretation.

After that, the first step is sampling, where the sample is obtained from the object to be analysed, collected in such a way that it represents the original object (e.g. if the metal concentration in a whole organ of a mammal should be quantified). Regarding tissue-specificity of heavy metal deposition [cf. 19] the choice between dissecting the organism or using the whole body is of great importance. Because of their size invertebrates, apart from gastropods and mussels, are mainly used in whole. To prevent bias only studies with the same body parts analysed should be compared.

All factors which (can) result in differences in the elemental concentrations must be taken in account, e.g. contamination by using metalic scoopfuls for soil samples.

The next step is the sample preservation, which is quite important, as there is usually a delay between sample collection and analysis. Sample preservation should ensure that the sample retains its physical and chemical characteristics so, that the analysis truly represents the object under study. Common preservation methods are drying, freezing, addition of preservatives such as alcohol or formalin, or of stabilizing agents (e.g. acids). Drying of the samples before digestion is necessary because concentration references to dry weight and there are differences in water content of the samples according to the sample preservation [20]. However drying must be exercised with caution as some metals, like Hg are volatile at temperatures below 120 °C. It should be taken into account, that preserving a sample using preservation solutions might change the properties of the sample (moisture content, extraction of analytes) [20,21].

Sample preparation is the next and last step prior to analysis, as most samples are not ready for direct introduction into analytical instruments. The analytes have to be released from the matrix, resulting in a clear solution, for example by digestion of the sample, which can then be analysed.

Digestion Methods

For metal analysis usually liquid samples are required. Common sample preparation methods are wet digestion in acid solution, dry ashing, and extraction of the analytes from the sample without total matrix destruction.

Wet Digestion

Commonly used methods for dissolving samples for metal analysis are digestion in an open container, digestion in a pressurized, sealed container, and microwave assisted decomposition. Usually the samples are prepared by digesting the matrix in a strong acid. To destroy organic matrices oxidizing agents are used. This leads to a clear digest solution. Nitric acid is usually applied, because there is no chance of forming insoluble salts as might happen with HCl or H_2SO_4. Hydrogen peroxide may be added to increase the oxidizing power of the digestion mixture. Digestion in an open container is more vulnerable to contamination of the sample and to losses of analytes due to the high volatility of several metals than the other wet digestion methods. The advantage of digestion in an open container is that large quantities of samples can be digested, however due to the low weight of invertebrates, this point is not necessary to be considered in this chapter.

Microwave assisted digestion in a closed system has several advantages over open container dissolution methods. This containers, made of strong high temperature resistent polymers, usually do not contain metal contaminants. In comparison with an open vessel decompostion step, a microwave assisted digestion procedure is less time consuming. The sealed container eliminates the possibility of airborne dust contamination and reduces evaporation, so that less digestion solution is required. Due to the higher pressures in the closed vessels temperatures above the boiling point of HNO_3 can be reached. Therefore oxidising ability of the acid as well as reaction rate are enhanced. In addition to that losses of volatile metal species can be eliminated if digestion solution is cooled down to room temperature before opening the vessels. The disadvantage of microwave assisted digestion is the low sample capacity and the danger of exploding when samples with high fat fraction are digested. Regarding low weight and fat fraction of invertebrates, microwave assisted digestion in closed vessels seems the best choice.

Dry Ashing

This is a relatively simple method of removing the organic matter and can be used for large amounts of samples. In the open vessel method, the sample is placed in a suitable crucible (made of silica, porcelain, or platinum) and heated in a muffle furnace. Major disadvantages of this method are possible loss of the analyte by volatilization, the contamination of the sample by airborne dust and the irreversible sorption of the analyte into the walls of the vessel.

Sample Slurries for Direct Anaylsis

Slurries are distributions of fine particles in a liquid and can be analysed by graphite furnace atomic absorption spectrometry (GFAAS). An advantage of this methods is the simple sample preparation, so that the risk of contamination is low and in addition no aggressive reagents are needed. The sample particle size however is critical, particles should be 2 μm or smaller. It is also important to keep the sample well homogenised during analysis.

Prevention of sample contamination from the reagents during the digestion process is required. In addition to contamination of the samples, there is also the possibility of losses of analytes during the preparation step: metal ions may adsorb on surfaces, especially on glass. Silica vessels are less likely to adsorb metals. Additionally some metals might be absorbed when the sample is heated in platinum crucibles, such as Fe, Hg, Au, and Pd. Adsorbed material is not only lost to the following analysis, but may also contaminate further samples prepared in these vessels. Volatile elements, such as Cd, Pb, Hg, Se, and Zn are lost when dry ashed, even at temperatures below 500°C. Remarkable is that the volatility depends on the matrix and the form of the metal.

Concluding it can be said that the kind of sample preparation depends on the sample, the matrix, and the analytes concentration. For instance, trace analysis requires more stringent sample preparation than major component analysis. Once the sample preparation is completed, the analysis is carried out e.g. by electrothermal or flame AAS. AAS-analysis of As and Se requires an hydrid generating system whereby a different sample preparation is necessary which is not considered in this chapter. In addition to that for toxicological studies especially of As, Cr and Hg the chemical form (speciation) of the analyte is of primary importance because the differences in toxicity, mobility, bioavailability, and bioaccumulation due to metal species.

Metals in terrestrial invertebrates have often been determined using different sampling and preparation methods. Various authors analysed invertebrates that had defecated [9,13,16,17]. In other cases, without prior defecation, animals were dissected and the tissues analysed separately [12,22,23]. Pitfalls mostly yielded dead specimens which, most likely, had not defecated. Traps with formalin [24-27] as well as pitfalls filled with a solution of ethanol [10,18] were used for sampling in former studies. Metal concentrations in formalin-fixed materials may differ from those found in fresh tissue, as e.g. Cu is eluted by the formalin [21]. Many authors did not indicate details of sampling and preparation methods, especially about how the animals were cleaned or dried [13,17,18]. Metal contamination was related to adsorption of metal dust on the hairs of spiders [28,29] indicating the importance of externally cleaning the invertebrates.

The final metal concentrations found in animals are the result of complex ecological relationships (e.g. life-history) and physiological processes (e.g. essential metal needs or toxic metal kinetics). Many variables influence the background level of metal concentrations, with usually strong inter-individual differences in natural populations. This background is superimposed by the foreground of variations due to decisions made by the investigator and by details of the methodology applied. Foreground variations may be tolerable as long as they are small and non-significant with respect to the stronger background variations. For such an evaluation, the effects of methodology must be tested using material taken *in situ*. This chapter examines differences in metal concentrations between diverse treatments in order to obtain further basic information for the design of standardised test preparation methods for samples taken from natural populations, particularly invertebrates, living in urban environments.

PRACTICAL APPROACH

Materials and Methods

Two areas in northern districts of the city of Vienna (Austria) were selected for sampling of gastropods (*Xerolenta obvia*), oligochaetes (*Lumbricus terrestris*), isopods (*Armadillidium vulgare* and *Trachelipus rathkei*) and carabids (*Harpalus rubripes* and *Calathus fuscipes*): Spitzau and Rendezvousberg were previously used as illegal waste dumps and are now ruderal 'wilderness' areas marked as 'suspected contamination areas' in the official register by the local authorities. During two weeks in May/June and in August/September 1997, the invertebrates (oligochaetes, gastropods, isopods, and carabids) were manually collected every second day. For the anecic oligochaete *Lumbricus terrestris*, a solution of 0.2-0.4% formalin was poured into a plastic frame marking 1953 cm^2 soil surface (figure 2); within approximately half an hour the lumbricids emerged, were manually collected and immediately washed with deionated water.

Figure 2: Plastic frame for sampling earthworms

Additional collections of isopods and carabids were made with pitfalls filled with a 3-4% formalin solution in deionated water (variations due to rainfall and evaporation). The pitfalls were set up over the two-week periods mentioned above. Invertebrates were transported to the laboratory in cooled Cd-free plastics boxes.

Standard treatment was defined by the following time-series: (1) manual collection, (2) defecation, (3) fixation by freezing at –18°C, (4) final species determination, sexing and sorting to size, (5) ultrasonic-cleaning (oligochaetes, isopods, carabids) or separation from shell (gastropods), (6) freeze-drying, (7) microwave assisted wet digestion with HNO_3 and (8) AAS analysis. Subsamples for the diverse treatments (see figures 3-6) were constituted by sorting the individuals to groups of about equal body size. This was not necessary for carabids, which showed very small size differences between imagines of a given species and sex. Alternative treatments are given below by their differences from the standard.

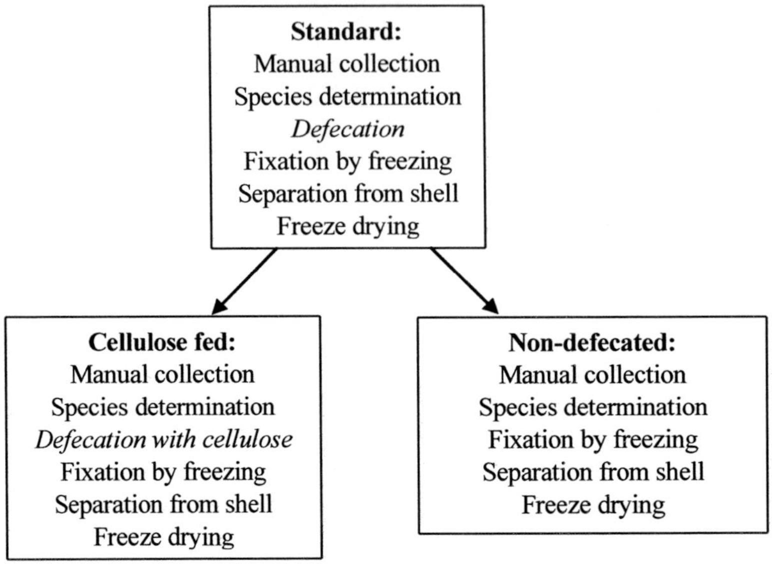

Figure 3: Subsamples of gastropods *(Xerolenta obvia)* for different methods of treatment

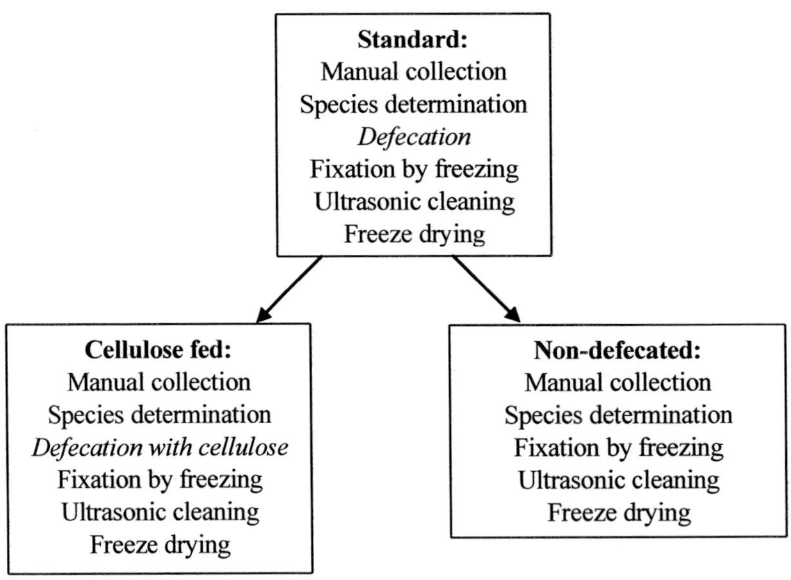

Figure 4: Subsamples of earthworms *(Lumbricus terrestris)* for different methods of treatment

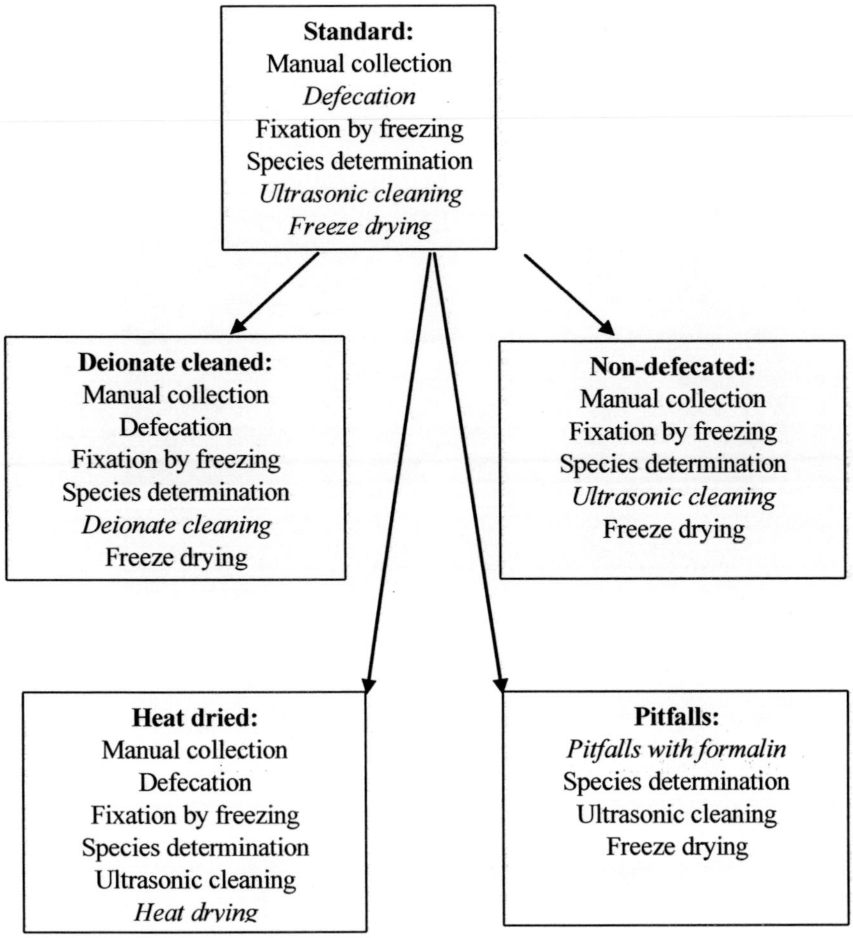

Figure 5: Subsamples of isopods *(Armadillidium vulgare* and *Trachelipus rathkei)* for different methods of treatment

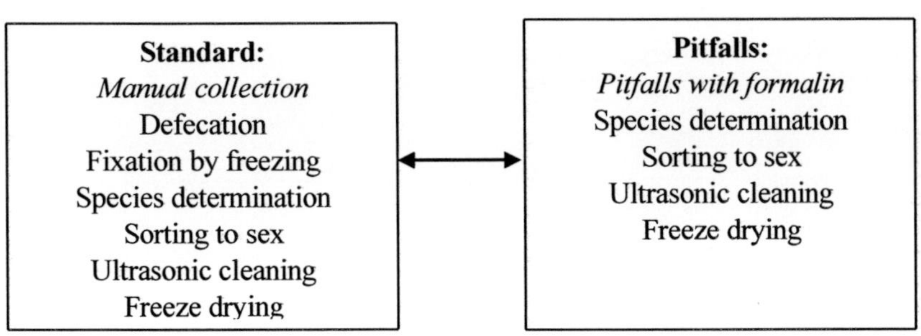

Figure 6: Subsamples of carabids *(Harpalus rubripes* and *Calathus fuscipes)* for different methods of treatment

Pitfall samples (invertebrates collected in a 3-4% formalin solution) inserted at point (4) into the series of the standard treatment. For defecation (standard treatment), invertebrates were kept in plastic boxes in an incubator at 18°C for four (gastropods) or six days (oligochaetes, isopods, and carabids). Pages of cellulose were added to the carabids to allow them to hole up. The boxes were cleaned every day and the faeces removed. Isopods were kept on grating in order to prevent coprophagy (figure 7). Petri dishes with deionated water were offered to them.

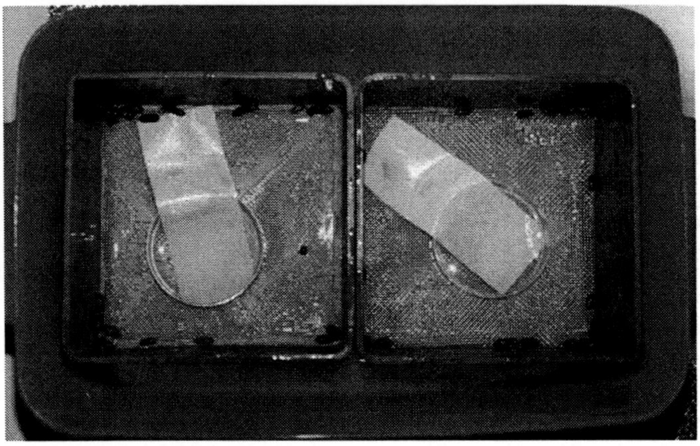

Figure 7: Grating for defecation of isopods

'Cellulose-fed': Oligochaetes and gastropods were allowed to defecate as in the standard treatment, but gut passage was accelerated by feeding small pieces of cellulose-strips soaked in deionated water (figure 8). Zn, Cu, Cd and Pb levels were below the detection limit in the cellulose-strips.

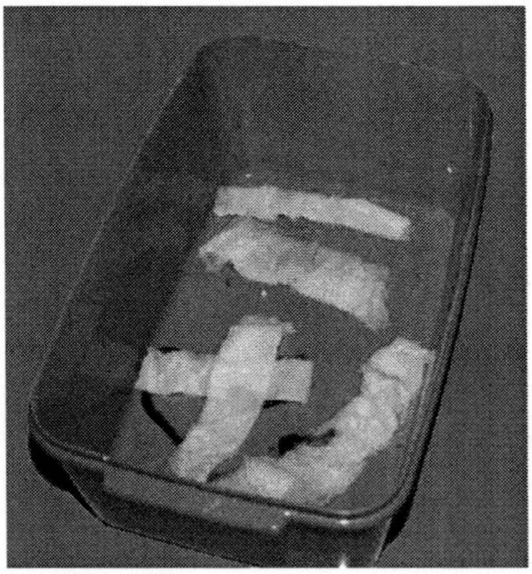

Figure 8: Earthworms defecated with cellulose as an accelerator of the gut passage

'Non-defecated': Invertebrates were manually collected and fixed by deep-freezing upon return to the laboratory [point (2) of the time-series of the standard treatment was missed out].

For ultrasonic-cleaning (standard treatment), the invertebrates were put in an ultrasonic bath for 15 min. Deionate-cleaning: Part of the isopod samples was briefly washed with deionated water instead of ultrasonic-cleaning. Freeze-drying (standard treatment): Samples of gastropods and carabids were freeze-dried at –48°C for 24 h, oligochaetes and isopods for 48 h until constance of weight. Heat-drying: Part of the isopod samples was dried for 48 h at

105°C instead of freeze-drying until constance of weight. Dried samples were stored over silicagel in an exsiccator until subsequent digestion.

For microwave assisted digestion, isopods, carabids and the soft bodies of gastropods were transferred into closed Teflon vessels with 3 ml of 65% HNO_3 (suprapure, Sigma Austria) or with 5 ml in the case of oligochaetes (table 1).

Table 1: Program of the microwave assisted digestion of the samples

Step	*Time (in min:sec)*	*Power (in Watt)*
1	03:00	250
2	00:30	0
3	05:00	250
4	05:00	400
5	00:30	0
6	02:00	400
7	00:30	0
8	02:00	500
9	00:30	500
Ventilation	03:00	

After cooling, the samples were rinsed with deionated water, filled up to a defined volume and stored in small PE-bottles.

Regarding metal analysis, Zn and Cu were measured with acetylene FAAS, Cd and Pb by GFAAS using a Polarized Zeeman Atomic Absorption Spectrophotometer (HITACHI Z 8200) (tables 2-5). All samples were measured twice. The procedure was repeated if the coefficient of variation was above 10%. Concentrations were expressed in $\mu g\ g^{-1}$ for Zn and Cu, as well as in $ng\ g^{-1}$ for Cd and Pb, using dry weights measured to 0.1 mg. Detection limits in the test solution were 4 $\mu g\ ml^{-1}$ for Zn, 30 $\mu g\ ml^{-1}$ for Cu, 0.25 $ng\ ml^{-1}$ for Cd, and 5 $ng\ ml^{-1}$ for Pb. Solutions of HNO_3 without samples prepared in the same way were used as blanks.

Table 2: Measurement conditions for FAAS (Zn)

Wavelength	213.8 nm
Flame	Air-C_2H_2
Burner Height	7.5 mm

Table 3: Measurement conditions for FAAS (Cu)

Wavelength	324.8 nm
Flame	Air-C_2H_2
Burner Height	7.5 mm

Table 4: Measurement conditions for GFAAS (Cd)

Wavelength	228.8 nm			
Injection Volume	27 µl			
Matrix Modifier	No			
Temperature Program				
Temperature (°C)			Time (sec)	
Stage	Start	End	Ramp	Hold
Drying	60	90	20	5
Drying	90	120	20	5
Drying	120	140	10	5
Ashing	150	250		5
Ashing	300	300		30
Atomization	1500	1500		5
Cleaning	1800	1800		8

Table 5: Measurement conditions for GFAAS (Pb)

Wavelength	283.3 nm			
Injection Volume	20 µl			
Matrix Modifier	Palladium (10 µl)			
Temperature Program				
Temperature (°C)			Time (sec)	
Stage	Start	End	Ramp	Hold
Drying	60	100	50	10
Drying	100	140	40	10
Drying	400	400		10
Ashing	600	600		10
Ashing	800	800		10
Atomization	2000	2000		5
Cleaning	2000	2200		10

Lyophilized bovine liver was used as reference material (BCR No. 185/1195; European Community Bureau of References, Belgium). The recoveries of all metals were above 91%.

Statistics were computed with SPSS 7.5 (SPSS Corp., USA). Data were normally distributed (Kolmogorov-Smirnov-test). Single-criterion ANOVA was used to evaluate differences related to the factor 'sample-treatment'. Homogeneity of variances was evaluated with the Levene-test. The Scheffé-test was applied when variances were homogeneous, elsewhere the Tamhane-T2-test was used. Significance was tested at $P \leq 0.05$ level.

RESULTS

Gastropods

The heavy metal concentrations recorded for the snail *Xerolenta obvia* at Spitzau and Rendezvousberg are given in table 6. Single-criterion ANOVA showed no significant differences for the Zn and Cu concentrations with regard to the defecation treatments tested. No significant differences were found between the defecation treatments in the samples from Rendezvousberg for Cd and Pb concentrations. At Spitzau, however, the 'non-defecated' group showed significantly lower Cd concentrations than the (defecated) standard group; similarly, Pb concentrations in the 'cellulose-fed' group were significantly lower compared to the (defecated) standard group. Coefficients of variation of metal concentrations ranked on the average Cu ≅ Zn ≅ Cd < Pb.

Table 6: Metal concentrations (mean ± SD) in *Xerolenta obvia* according to different collecting sites, sampling methods and sample treatments. Sample locations and periods in parentheses: Spitzau (S), Rendezvousberg (R), May/June (1) or Aug./Sept. (2) 1997. Significant differences (single-criterion ANOVA; p < 0.05) from standard treatment are denoted in bold

Method of treatment	*N*	*Location/ period*	*Zn (μg g⁻¹)*	*Cu (μg g⁻¹)*	*Cd (ng g⁻¹)*	*Pb (ng g⁻¹)*
Standard	20	S2	404±117	64±28	2464±767	2878±1838
Cellulose-fed	20	S2	329±129	54±24	1958±683	**1742±936**
Non-defecated	21	S2	332±122	68±22	**1812±436**	2060±1122
Standard	20	R2	323±144	80±15	1795±385	3402±1540
Cellulose-fed	20	R2	265±106	67±21	1620±559	2481±1510
Non-defecated	20	R2	238±97	68±17	1862±742	2947±1712

Earthworms

Lumbricus terrestris was sampled only at Spitzau. The metal concentrations following different defecation treatments are listed in table 7. Compared with the (defecated) standard group, significantly higher concentrations of Cu and Pb and lower concentrations of Zn and Cd were found in the 'non-defecated' animals. No significant effects were observed for the 'cellulose-fed' group. Coefficients of variation ranked on the average Cu < Zn < Cd < Pb.

Isopods

Metal concentrations for *Armadillidium vulgare* and *Trachelipus rathkei* are given in table 8. Compared with manual (standard) collections, isopods caught with formalin-pitfalls showed significantly lower Cu but higher Zn, Cd and Pb concentrations in some of the experiments. Compared with 'defecated' (standard) animals, 'non-defecated' *A. vulgare* had significantly higher Pb concentrations. Similarly, compared with ultrasonic-cleaned (standard) animals, 'deionate-cleaned' *A. vulgare* showed significantly higher Cd concentrations in one of the experiments. Coefficients of variation for *A. vulgare* ranked on the average Cu ≅ Zn < Cd < Pb; the corresponding ranking for *T. rathkei* was Zn ≅ Cu < Pb < Cd.

Table 7: Metal concentrations (mean ± SD) in *Lumbricus terrestris* according to different collecting sites, sampling methods and sample treatments. Sample locations and periods in parentheses: Spitzau (S), Rendezvousberg (R), May/June (1) or Aug./Sept. (2) 1997. Significant differences (single-criterion ANOVA; p < 0.05) from standard treatment are denoted in bold

Method of treatment	*N*	*Location/ period*	*Zn (µg g^{-1})*	*Cu (µg g^{-1})*	*Cd (ng g^{-1})*	*Pb (ng g^{-1})*
Standard	17	S2	327±81	12±2	4212±1342	1070±732
Cellulose-fed	14	S2	389±170	12±2	5467±3727	1043±963
Non-defecated	13	S2	**235±54**	**15±2**	**3023±1151**	**2664±1829**

Table 8: Metal concentrations (mean ± SD) in *Armadillidium vulgare* and *Trachelipus rathkei* according to different collecting sites, sampling methods and sample treatments. Sample locations and periods in parentheses: Spitzau (S), Rendezvousberg (R), May/June (1) or Aug./Sept. (2) 1997. Significant differences (single-criterion ANOVA; p < 0.05) from standard treatment are denoted in bold

Method of treatment	*N*	*Location/ period*	*Zn (µg g^{-1})*	*Cu (µg g^{-1})*	*Cd (ng g^{-1})*	*Pb (ng g^{-1})*
Armadillidium vulgare						
Standard	21	S1	160±57	247±89	299±100	1324±657
Deionate-cleaned	9	S1	141±64	217±59	288±62	735±495
Heat-dried	20	S1	168±43	267±62	234±108	944±727
Pitfalls	20	S1	165±89	181±108	265±148	2093±1274
Standard	20	S2	170±47	241±55	256±122	756±366
Deionate-cleaned	20	S2	156±37	230±61	272±79	1139±687
Heat-dried	20	S2	162±67	228±48	217±119	693±378
Non-defecated	20	S2	147±32	196±56	236±74	**1330±719**
Pitfalls	20	S2	160±54	**140±29**	289±62	**1945±710**
Standard	22	R1	156±49	189±53	197±74	1217±664
Deionate-cleaned	20	R1	141±26	190±33	**302±104**	1248±519
Heat-dried	22	R1	140±31	199±40	187±64	1011±620
Pitfalls	13	R1	**190±44**	164±47	**464±178**	**3114±1137**
Trachelipus rathkei						
Standard	20	R2	168±49	264±92	315±263	1139±561
Heat-dried	20	R2	147±52	205±79	523±387	1636±1127
Pitfalls	20	R2	181±38	**183±50**	**953±752**	**2099±579**

Carabids

Metal concentrations for *Harpalus rubripes* and *Calathus fuscipes* are listed in table 9. Significant differences in Zn, Cu and Cd concentrations were found between the sexes in both species. The female bodies contained more Zn and Cd whereas the males showed a higher Cu concentration (about twice that in females). No more (significant) differences could be found for *C. fuscipes*. Compared with manual (standard) collections, *H. rubripes* caught with formalin-pitfalls showed significantly higher Zn (both sexes) and Cd (males only) concentrations in some of the experiments. Coefficients of variation for both sexes of *H. rubripes* ranked on the average Zn << Cu < Cd < Pb; this ranking was Zn < Cu < Pb ≅ Cd for *C. fuscipes*.

Table 9: Metal concentrations (mean ± SD) in *Harpalus rubripes* and *Calathus fuscipes* according to different collecting sites, sampling methods and sample treatments. Sample locations and periods in parentheses: Spitzau (S), Rendezvousberg (R), May/June (1) or Aug./Sept. (2) 1997. Significant differences (single-criterion ANOVA; p < 0.05) from standard treatment are denoted in bold, below detection limit by 'n.d.'

Method of treatment	*N*	*Location/ period*	*Zn (µg g⁻¹)*	*Cu (µg g⁻¹)*	*Cd (ng g⁻¹)*	*Pb (ng g⁻¹)*
Harpalus rubripes						
Standard,females	15	S1	129±8	12±4	58±104	n.d.
Pitfalls, females	9	S1	**151±17**	12±4	63±43	111±82
Standard,males	25	S1	105±7	35±11	20±18	n.d.
Pitfalls, males	7	S1	130±21	28±10	**43±17**	116±142
Standard,females	5	R1	117±11	20±9	49±52	n.d.
Pitfalls, females	6	R1	140±23	12±7	51±14	89±119
Standard,males	12	R1	101±10	33±10	17±17	n.d.
Pitfalls, males	5	R1	**128±5**	35±25	47±18	76±102
Calathus fuscipes						
Standard,females	18	S2	128±16	25±13	205±117	140±76
Pitfalls, females	10	S2	170±48	38±9	630±374	118±210
Standard,males	25	S2	120±10	56±18	87±55	161±100
Pitfalls, males	10	S2	116±17	51±7	191±292	n.d.
Standard,females	21	R2	131±16	19±14	151±212	201±95
Pitfalls, females	10	R2	113±36	23±5	194±225	n.d.
Standard,males	14	R2	120±8	56±16	29±17	197±49
Pitfalls, males	10	R2	133±22	56±12	217±218	n.d.

Discussion

Variability of metal concentrations: Measurements of Zn, Cu, Cd and Pb in the bodies of selected invertebrates at urban sites were not homogeneous with respect to the coefficient of variation (calculated directly from standard deviations and means in tables 6-9). The potentially toxic metals Cd and Pb generally showed larger inter-individual variability than the biologically essential metals Zn and Cu. This might be caused by regulation mechanisms of essential trace elements by the invertebrates. It is important to note that higher relative differences are required in Cd and Pb concentrations between the groups than in Zn and Cu´s in order to obtain a certain significance level at given numbers of specimens. Therefore sample size calculation before sampling according to a preliminary test for estimating the standard deviation is of great importance to minimize the statistic β-error. For Zn-concentrations in the carabid *Harpalus rubripes* (table 9), coefficients of variation were only 4-17 %, which is exceptionally low for natural populations of invertebrates in urban environments. The comparatively small variations of Cu in snails and isopods coincide with the role of Cu as a component of the respiratory pigment (hemocyanin) in these animals. A possible regulation of Cu was proposed for isopods before [30].

Gastropods

In this chapter, the concentrations of metals varied considerably between individuals. With regard to the different defecation treatments significantly higher Cd concentrations were detected in animals which had defecated compared to those which had not; 'cellulose-fed' snails displayed lower Pb-concentrations than normally defecated individuals in one of the two experiments (table 6). This increase in the Cd concentration in defecated animals might be due to the lower body weight correlating with defecation as there is more metal per unit. Cd is not excreted by feces and might accumulate in the gastropod. For detoxification it can be bound in certain tissues (e.g. in the midgut gland) to metallothioneins [31-33]. Most authors allowed the gastropods to defecate before measurements; exceptions are Pavel et al. [34] as well as Bengtsson et al. [29], who used pitfalls with ethylene glycol. It was shown that Cd concentrations in the snail *Cepaea hortensis* were correlated with age, at which the animals had defecated before measurement [13]. The hepatopancreas (similar to the chloragogen tissue of earthworms) is involved in the storage of Zn and the detoxification of Pb [35]. The dissected hepatopancreas of adult snails was already suggested to be used for monitoring purposes instead of the whole animals [36]. Dallinger [32] described strategies of metal detoxification of *Helix pomatia* by cellular compartmentalisation of Pb in cells of the midgut gland. Also other metals apart from Cd (e.g. Cu) can be detoxificated by binding to metallothioneins. Different metallothionein-isoforms of midgut gland (induceable by Cd) and mantle (reduced by physiological stress) in *Helix pomatia* and *Cepaea hortensis* might be used as specific biomarkers for Cd poisoning [31,33,37-39].

Earthworms

As soil animals oligochaetes are exposed to generally high concentrations of heavy metals, and especially the soil surface may be highly contaminated [40,41]. *Lumbricus terrestris* usually collects food from the soil surface and moves down the soil for feeding, digestion and excretion. Therefore metal exposure of *L. terrestris* is not equal to metal consumption. Furthermore it should be considered, when attempting earthworms for bioindicators, that the calcium content and pH-value of the soil are of major relevance for the bioavailability of metals [4]. The data of this chapter show that different sample treatments may highly influence the heavy metal concentrations measured in the earthworms. In particular, defecated earthworms showed significantly different concentrations of Zn, Cu, Cd and Pb compared to non-defecated animals. For this comparison, note that the intestinal content may contribute up to more than half of the body dry weight. Similar as found for Cd in gastropods, the higher accumulation of Cd and Zn in defecated animals might be due to the lower bodyweight after defecation. With few exceptions, most literature data are for defecated earthworms. However in former studies formalin-pitfalls were used for sampling, inevitably collecting animals at different states of defecation [26]; this method, therefore appears to be impractical for metal analysis. It was found that earthworms strongly accumulate Zn, whereby animals had defecated and were freeze-dried before measurement [42]. Pb and Cd are also accumulated to high concentrations by earthworms [42-46]. Morgan et al. [47] showed that Cd, Pb and Zn in *Lumbricus rubellus* is primarily accumulated in the posterior alimentary canal, thus may represent a detoxification strategy based on accumulative immobilisation. *Lumbricus terrestris* accumulates Cd more than other species of earthworms [45]. The Pb concentrations also differ between species [45]. Pb is stored and detoxified in chloragogen tissue around the gut [35]. For Cu no evidence of sequestration was found [47].

Isopods

In previous studies sampling for isopods varied between manually collection and pitfall trapping. Some authors allowed the isopods to defecate before measurement [46,48] in contrary to Hunter et al. [27] who used formalin-pitfalls. At least in some experiments of this chapter, pitfall-collected isopods showed significantly higher concentrations of Zn, Cd and Pb, but lower concentrations of Cu compared to manually collected individuals. Cu is eluted from tissues due to formalin fixation [21]. In one of the experiments of this chapter, non-defecated isopods showed a significantly higher Pb-concentration compared to defecated ones. Lower concentrations of Zn, Cu, Cd and Pb were also found in former studies in specimens with empty gut or during moulting [23]. It was shown that Pb-concentration varied between different isopod species and was not above the respective concentrations in the soil [46]. In addition strong species-specific differences in Zn and Cd accumulation were found in terrestrial isopods (*Oniscus asellus* and *Porcellio scaber*) [22]. Similar as in snails Cd can be bound to proteins in the cytosol of the hepatopancreas in isopods for detoxification purpose [23].

Coleoptera

Carabids usually accumulate metals at low levels [49,50] but appear to be more sensitive than some herbivorous arthropods [29]. For example decreasing numbers of carabid species and individuals were found along immission gradients of a busy road [51]. In some experiments of this chapter, significantly higher Zn or Cd-concentrations were found in specimens collected with formalin-pitfalls versus manually collected animals. Gut content may be an important factor for the total metal burden of carabids [11,52], as the gut and the Malpighian tubules are important storage organs of metals [50,53]. Rittner [50] found 85 % of the initial Cd-concentrations in the gut of carabids after two days of defecation. This chapter showed significant sex-specific differences for Zn, Cu, and Cd-concentrations of *Harpalus rubripes* and *Calathus fuscipes* (table 9). This was also reported for Cu, Cd and Pb-concentrations in pitfall-trapped beetles (*Aptinus bombarda* and *Abax parallelepipedus*) [40]. In addition to that, life stage of the coleoptera seems to be of great importance for metal accumulation. Hunter et al. [27], using beetles from formalin-pitfalls, and Carter [54], using defecated ones, found higher levels of Cd in larvae than in imagines. Cd-concentrations in *Pterostichus niger* were lower during moulting or pupation [55]. This might also be a strategy for detoxification of heavy metals. Correlations between Cd-concentration and body size of carabids were found in a former study whereby animals that had been caught alive but which had not defecated were used [56]. In this chapter only adult beetles of similar size were used for metal determination.

Sampling Methods

The use of formalin-pitfalls versus manual collections showed a strong effect on metal concentrations in isopods, and to a lesser extent also in carabids (tables 8, 9). Many authors collected invertebrates by pitfall trapping and consequently analysed specimens with (partly) filled gut [10,18,24-27]. The different methods of treatment may have contributed to certain inconsistencies in the literature: Roberts et al. [24] as well as Hunter et al. [26] used formalin-pitfalls and concluded that Cd may accumulate along terrestrial food chains. In contrast, van Straalen et al. [11] showed that the physiology of different animal species may be much more important for metal concentrations than biomagnification. This was confirmed by Heck et al. [46].

Defecation

A major methodological problem stems from the fact that animals collected in the field often show strong inter-individual or temporal differences in gut-filling. In all four metals examined, defecated animals, particularly earthworms, differed significantly from those which had not defecated before (table 7). The importance of allowing the animals to defecate before analysis is also underlined by the detoxification of heavy metals via intestinal excretion [11,52]. The literature data show a general trend toward letting the animals defecate before analysis [9,13,16,17]. In those cases where invertebrates with filled gut were analysed, the validity of the data on incorporated metals appears to be questionable.

Acceleration of Gut Passage

Upon defecation in the laboratory, several snail and earthworm individuals did not entirely eject the initial gut content. Complete ejections were obtained by feeding small pieces of cellulose (see 'Materials and Methods'). For the resulting metal concentrations in earthworms, however, this cellulose treatment did not appear to be important (table 7), at least compared with the background of high interindividual variation. The same was observed in snails, except that Pb-concentrations were significantly lower in one of the cellulose-feeding experiments (table 6). The cellulose-feeding treatment seems to be too ineffective for a general recommendation of this method, especially because feeding may introduce additional sources of individual variation, contamination and possible bias.

Fixation

The use of formalin influences heavy metal concentrations by elution, as has been shown for the entire body of isopods and for autopsied human tissue [21]. This chapter did not directly test this well-known fact, although formalin-fixation most likely contributed to the difference between manually collected, freeze-fixed ('standard' in table 8) animals versus those collected by formalin-pitfalls. According to own experiments, fixation with solutions of formalin or ethanol induced strong slime exudation in snails and earthworms. This could be avoided by freeze-fixing.

Species Determination, Sexing and Sizing

Animals should be determined to species level before measurement because of frequently found species-specific differences in heavy metal accumulation [45]. As also shown previously [40], this chapter illustrates the great importance of sorting coleopteras due to their sex. In addition, there are differences according to moult stages and pupation [27,54,55]. In this chapter only adult specimens were used, and potential effects of body size and age were compensated by dividing the samples into groups of about equally sized animals for each of the experimental treatments.

Cleaning

One of three experiments with isopods yielded a significantly higher Cd-concentration in deionate-washed versus ultrasonic-cleaned specimens. Based on this result and others on invertebrates [28,29], ultrasonic cleaning is recommended for arthropods, provided that the material is not damaged by ultrasonic.

Drying

No significant differences between heat-dried and freeze-dried isopods were detected for Zn, Cu, Cd, and Pb (table 8) as these four metals are relatively non-volatile. Therefore, the present results may not be transferable to more volatile elements such as Hg and As.

CONCLUSION

Metal concentrations tend to differ strongly between individuals from natural populations of invertebrates. Standard deviations are often very high. Therefore necessity of sample size calculation has to be pointed out, especially if non essential heavy metals like Cd or Pb are measured. The use of different preparation methods bears the risk of introducing additional bias. Due to the results of this chapter we recommend that all invertebrates should be collected *in vivo*, be allowed to defecate, be freeze-fixed and (at least the arthropods) ultrasonic-cleaned, determined to species level and in certain cases to sex and (age) stage, and then finally sorted by size before further sample preparation and analysis. For sample preparation microwave assisted digestion is suitable as the disadvantages of this method can be disregarded if invertebrates are used. Special attention has to be turned on the temperature of the drying steps of AAS analysis and the digestion method if volatile elements should be examined. It has to be underlined that every single step of sampling and sample preparation bears the risk to contaminate the sample or to loose volatile analytes. Additionally every single step of sample treatment enhances the standard deviation (Gauss'law of error propagation). Needless steps should therefore be avoided. Vessels used in sample preparation steps should be cleaned before use with acid (e.g. 7 N HNO_3) and rinsed three times with double distilled water. Contact with materials which might include the analyte (e.g. in scoopfuls or tweezers) has to be strictly avoided during sampling and sample preparation steps. Metal concentrations in the reagences (like acids) or additional contact materials (like in this chapter the cellulose) should be determined as well. As a minimum requirement, especially when data from past and present investigations are to be compared, the use of comparable sampling and preparation methods is recommended in order to avoid possible bias in the results and subsequent interpretation.

ACKNOWLEDGEMENTS

Many thanks to Elsevier Science for the right to extent the article [57] into book-length form. The practical approach was supported by the 'Hochschuljubiläumsstiftung der Stadt Wien' (project no. H-00075/96) and by funds of the Austrian National Bank (no. 6260/4).

BIBLIOGRAPHY

[1] U. Arndt, W. Nobel, B. Schweizer, Bioindikatoren. Möglichkeiten, Grenzen und neue Erkenntnisse pp. 396, *Ulmer Verlag*; 1987.

[2] H. Parlar, D. Angerhöfer, Chemische Ökotoxikologie pp. 386, *Springer Verlag*; 1995.

[3] R. Schubert, Bioindikation in terrestrischen Ökosystemen pp. 338, *Fischer Verlag*; 1991.

[4] S.P. Hopkin, Ecophysiology of metals in terrestrial invertebrates pp. 366, *Elsevier Appl. Science*; 1989.

[5] W. Wieser, *Am. Zool.* 8 (1968) 495-506.

[6] A. Nicolaidou, J.A. Nott, *Mar. Pollut. Bull.* 36 (1998) 360-365.

[7] S.G. Kang, M.S. Choi, I.S. Oh, D.A. Wright, C.H. Koh, *Sci. Total Environ.* 234 (1999) 127-137.

[8] T. Koeckritz, U. Irmler, P. Weppen, *J. Plant Nutr. Soil Sci.* 162 (1999) 477-482.

[9] K. Grodzinska, B. Godzik, E. Darowska, B. Pawlowska, *Ekologia Polska* 35 (1987) 327-344.

[10] F.E. Giles, S.G. Middleton, *J.G. Grau, Environ. Entomol.* 2 (1973) 299-300.

[11] N.M. van Straalen, J. van Wensen, *Environ. Pollut.* 42 (1986) 209-221.

[12] R.S. Sohal, R.E. Lamb, *J. Insect Physiol.* 25 (1979) 119-124.

[13] P. Williamson, *Ambio* 8 (1979) 30-31.

[14] C. Gundacker, *Environ. Pollut.* 110 (2000) 61-71.

[15] S. Gintenreiter, J. Ortel, H.J. Nopp, *Arch. Environ. Contam. Toxicol.* 25 (1993) 55-61.

[16] N.M. van Straalen, T.B.A. Burghouts, M.J. Doornhof, G.M. Groot, M.P.M. Janssen, E.N.G. Joosse, J.H. van Meerendonk, J.P.J.J. Theeuwen, H.A. Verhoef, H.R. Zoomer, *J. Appl. Ecol.* 24 (1987) 953-968.

[17] E.N.G. Joosse, J.B. Buker, *Environ. Pollut.* 18 (1979) 235-240.

[18] C.L. Strojan, *Oikos* 31 (1978) 41-46.

[19] C. Gundacker, *Chemosphere* 38 (1999) 3339-3356.

[20] M. Zeiner, I. Steffan, *in preparation* (2004)

[21] N. Koizumi, F. Hatayama, K. Sumino, *Environ. Res.* 64 (1994) 192-198.

[22] S.P. Hopkin, *J. Appl. Ecol.* 27 (1990) 460-474.

[23] S.P. Hopkin, M.H. Martin, *Oecologia* 54 (1982) 227-232.

[24] R.D. Roberts, M.S. Johnson, *Environ. Pollut.* 16 (1978) 293-310.

[25] K.J. Wade, J.T. Flannagan, A. Currie, D.J. Curtis, *Environ. Pollut.* 1 (1980) 87-93.

[26] A. Hunter, M.S. Johnson, *Oikos* 38 (1982) 108-117.

[27] A. Hunter, M.S. Johnson, D.J. Thompson, *J. Appl. Ecol.* 24 (1987) 587-599.

[28] S. Gintenreiter, Schwermetallbelastung von Spinnen an zwei unterschiedlich exponierten Standorten *Master's thesis*, University of Vienna; 1990.

[29] G. Bengtsson, S. Rundgren, *Ambio* 13 (1984) 29-33.

[30] W. Wieser, L. Wiest, *Oecologia* 1 (1968) 38-48.

[31] R. Dallinger, B. Berger, P. Hunziker, J.H. Kagi, *Nature* 388 (1997) 237-238.

[32] R. Dallinger, in: R. Dallinger,P.S. Rainbow (Ed.), Ecotoxicology of metals in invertebrates, Strategies of metal detoxification in terrestrial invertebrates, *Lewis Publishers,* Boca Raton, USA, 1993, pp. 245-289.

[33] R. Dallinger, B. Lagg, M. Egg, R. Schipflinger, M. Chabicovsky, *Ecotoxicology* (2003) in press.

[34] J. Pavel, D. Povolny, *Sci. Total Environ. Suppl. Part 1* (1993) 449-452.

[35] H.O.W. Ernst, E. Joosse van Damme, Umweltbelastung durch Mineralstoffe-biologische Effekte *Fischer Verlag*; 1983.

[36] W.B. Rabitsch, Bull. *Environ. Contam. Toxicol.* 56 (1996) 734-741.

[37] R. Dallinger, M. Chabicovsky, B. Berger, *Environ. Toxicol. Chem.* (2003) in press.

[38] R. Dallinger, M. Chabicovsky, B. Lagg, R. Schipflinger, H.G. Weirich, B. Berger, *Environ. Toxicol. Chem.* (2003) in press.

[39] M. Chabicovsky, H. Niederstatter, R. Thaler, E. Hodl, W. Parson, W. Rossmanith, R. Dallinger, Toxicol. *Appl. Pharmacol.* 190 (2003) 25-36.

[40] W.B. Rabitsch, *Environ. Pollut.* 90 (1995) 221-237.

[41] P. Williamson, R.P. Evans, *Pedobiologia* 13 (1973) 16-21.

[42] R.I. van Hook, Bull. *Environ. Contam. Toxicol.* 12 (1974) 509-513.

[43] C.D. Gish, R.E. Christensen, *Environ. Sci. Technol.* 7 (1973) 1060-1062.

[44] M.P. Ireland, *Environ. Pollut.* 19 (1979) 201-206.

[45] C.D.J. Ash, D.L. Lee, *Environ. Pollut.* 22A (1980) 59-68.

[46] M. Heck, U. Rink, G. Weigmann, Z. Ökol. *Naturschutz* 4 (1995) 75-85.

[47] J.E. Morgan, A.J. Morgan, *Oecologia* 84 (1990) 559-566.

[48] W. Kratz, H. Gutke,G. Weigmann, *Acta Phythopath. Entom. Hung.* 22 (1987) 391-397.

[49] M.P.M. Janssen, R.F. Hogervorst, *Environ. Pollut.* 79 (1993) 181-189.

[50] P. Rittner, Organspezifische Verteilung von Blei und Cadmium in Insekten unterschiedlicher Trophieebenen *Master's thesis*, University of Ulm; 1991.

[51] R. Maurer, *Oecologia* 14 (1974) 327-351.

[52] L. Lindquist, M. Block, *Comp. Biochem. Physiol.* 111C (1995) 325-328.

[53] E. Hörth, In: Untersuchung zur Verteilung von Cadmium, Blei und Zink in den Organen der Schabe Periplaneta americana *Master's thesis*, University of Vienna; 1989.

[54] A. Carter, *Can. J. Zool.* 61 (1983) 2751-2757.

[55] L. Lindquist, M. Block, H. Tjälve, *Environ. Toxicol. Chem.* 14 (1995) 1195-1201.

[56] M.P.M. Janssen, J.J.M. Bedaux, *Neth. J. Zool.* 39 (1989) 194-207.

[57] B. Zodl, K.J. Wittmann, *Chemosphere* 52 (2003) 1095-1103.

In: Water Pollution: New Research
Editor: A. R. Burk, pp. 99-117
ISBN: 1-59454-393-3

Chapter 5

APPLICATION OF ADVANCED OXIDATION PROCESSES (AOPS) TO WASTEWATER TREATMENT. CASE STUDIES: DECOLOURIZATION OF TEXTILE EFFLUENTS, DETOXIFICATION OF OLIVE MILL EFFLUENT, TREATMENT OF DOMESTIC WASTEWATER

***Nuri Azbar*[1*], *Kadir Kestioglu*[2] *and Taner Yonar*[2]**
[1]Ege University, Bioengineering Department, 35100, Bornova - Izmir, Turkey
[2] Uludag University, Environmental Engineering Department, 16059, Gorukle, Bursa, Turkey

ABSTRACT

Advanced oxidation processes (AOPs) are emerging and promising technology both as an alternative treatment to conventional wastewater treatment methods and enhancement of current biological treatment methods especially dealing with highly toxic and low biodegradable wastes. In this paper, the various advanced oxidation processes and their common principles were briefly discussed. In addition, reactors and light sources used most frequently are given in summary. Finally, the application of AOPs to wastewater treatment focusing on some case studies (coloured textile effluent, olive mill effluent, domestic wastewater) is presented.

Key words: Advanced oxidation, textile effluent, olive mill waste, domestic wastewater.

[*]Corresponding author, E-mail: nuriazbar@yahoo.com; Fax : +90 – 232 - 388 49 55

INTRODUCTION

The permitted levels of even treated effluents have been vastly exceeded, causing such environmental contamination that our natural resources, whose auto-purifying capacity is already exhausted, cannot be used for certain uses and their characteristics have been altered. The main problem stems from waste forms such as phenols, pesticides, fertilizers, detergents, and other chemical products resulting from industry and agriculture, despite the fact that the population also plays an important role in environmental contamination. Biological degradation processes are often not efficient for industrial effluents with toxic and/or non-biodegradable organic substances. Therefore it is very clear that the search for the development of effective means of removing toxic or non-biodegradable compounds and new approach to degrade the xenobiotics to biocompatible materials is of interest to regulating authorities and a significant issue in the agenda of both developed and developing countries.

AOPs, which have been proven to be able to successfully destroy numerous organic and halogenated organic chemicals, have been shown to be powerful treatment alternatives for this purpose (Azbar et al., 2004; Beltran et al., 1993; Holf et al., 1997; Mokrini et al., 1997; Ku et al., 1998). In general, AOPs were applied for the disinfection of both drinking water and treated wastewater effluents. However, it has been proven that use of AOPs have a great advantage of improving the biodegradability, when combined with biological treatment methods. On the other hand, the AOPs show specific advantages over conventional treatment alternatives because it can eliminate non-biodegradable organic components and avoid deposit residues as a special waste. A wide variety of organic and explosive contaminants are susceptible to destruction by AOP including petroleum hydrocarbons, halogenated solvents, phenolic compounds, pentachlorophenol, pesticides, dioxin, glycols, polychlorinated biphenyles, explosives such as TNT, RDX, and HMX, creosote, Freon 113, vinyl chloride, benzene, toluene, ethylbenzene, and xylenes (BTEX), methyl tertiary butyl ether, cyanide, mixed organic/radioactive waste and other organics. Also, organisms such as *Salmonella* and *E.Coli* have reportedly been significantly reduced using AOP such as UV/oxidation.

A partial list of AOPs, which involve the in situ generation of highly potent chemical oxidant (hydroxyl radical), includes homogeneous ultraviolet irradiation (Bolton and Cater, 1994) (either direct irradiation of the contaminant or photolytic oxidation mediated by hydrogen peroxide (UV/H_2O_2) and/or ozone ($UV/H_2O_2/O_3$), heterogeneous photocatalysis using semiconductor catalysis (e.g., UV/TiO_2), electron-beam irradiation, X-ray or gamma-ray radiolysis non-thermal electric discharge, supercritical water oxidation and ultrasonic irradiation (sonolysis) or electrohydraulic cavitations. Widely different methods of activation, as well as oxidant generation are involved in these processes which are mostly electric-energy-driven and share the common denominator of hydroxyl radical chemistry for part of the pollutant removal. In this paper, discussion and examples will be focused on those of most used AOPs.

Principles of Advanced Oxidation Processes (AOPs)

The advanced oxidation processes (AOPs) are based on two main steps; the generation of free radicals and the oxidation reactions of the polluting molecules with these radicals as follows:

$$\text{AOP} \xrightarrow{} OH^{\bullet} \xrightarrow{\text{Pollutants}} CO_2 + H_2O + \text{Inorganic ions}$$

The hydroxyl radical is primarily responsible for the efficiency of the AOPs. Its high oxidation potential (2.8 V; Hong et al., 1996) in fact guarantees a rapid degradation of the polluting molecules. The life span of the $OH^{\bullet}$ radical is so short about 10 μs, which therefore dictates the production of this radical in situ through chemical or photochemical reactions. However, it is still a long time for the radical to show its oxidizing action and promote the formation of a chain reaction capable of producing other free radicals (Crittenden et al., 1999; Hogne, 1997). The kinetics with respect to hydroxyl radical concentrations and to the pollutant seems to be first order with the pseudo first order constant in the range of $1\text{-}10^{-4}$ s (Esplugas et al., 2002; Pera-Titus et al., 2004).

i. ***Ozone (O_3):*** Ozone is a powerful oxidizer having an oxidation potential of about 1.5 times that of chlorine gas. The two possible oxidizing actions, namely direct and radical way, are considered in an ozonation process. The direct way is due to the reaction between the ozone and the dissolved pollutant, and the radical way is based on reactions between the generated hydroxyl radicals in the ozone decomposition and the dissolved pollutant (Hoigne and Bader, 1983). At low pH, ozone exclusively reacts with compounds with specific functional groups through selective reactions such as electrophilic, nucleophilic or dipolar addition reactions (Langlais et al., 1991), on the other hand at basic conditions, ozone decomposes yielding hydroxyl radicals (Legrini et al., 1993) that reacts non-selectively with organic and inorganic compounds (Pera-Titus et al., 2004).Ozone decomposition to produce hydroxyl radical can be enhanced by the help of Mn^{+2}/oxalic acid (Andreozzi et al., 1992) at pH>4.0 at which Mn(III)-dioxalate and Mn(III)-trioxalate are formed. The formation of hydroxyl radicals proceeds as a result of a reaction between manganese complexes and ozone as follows (Andreozzi et al.,1995);

$$Mn(III)(AO^{2-})_n + O_3 + H^+ \longrightarrow Mn(II) + (n-1)(AO^{2-}) + 2CO_2 + O_2 + HO$$

ii. ***Ozone + Hydrogen Peroxide:*** The addition of hydrogen peroxide to the ozone aqueous solution will enhance the O_3 decomposition with formation of hydroxyl radicals. Interaction between the ozone and the hydrogen peroxide generates the hydroxyl radicals as described below (Adams et al., 1994; Glaze and Kang, 1989; Hoigne, 1998; Stachelin and Hoigne, 1982);

$$H_2O_2 + 2\,O_3 \longrightarrow 2\,OH^{\bullet} + 3\,O_2$$

The pH of solution is critical for the processs efficiency, since in the ozone decomposition mechanism the active species is the conjugate base HO_2^- whose concentration is strictly dependent upon pH. Addition of hydrogen peroxide to the aqueous O_3 solution at high pH conditions will result in higher production rates of hydroxyl radicals (Glaze an Kang, 1989). One advantage of this sytem is that it can be used in turbid or dark waters because it is a dark AOP that is not dependent on the UV transmissivity of the influent being treated. In addition, the adoption of the H_2O_2/O_3 process is fairly easy and do not require significant changes to the apparatus, since it is only necessary to add an H_2O_2 dosing system.

iii. ***UV radiation:*** This method is based on the absorption of energy supplied by UV radiation to the chemical compounds by reactant molecules (Legrini et al., 1983).

iv. ***UV radiation + Ozone***: Interaction between O_3 and the energy supplied by UV radiation (254 nm) results in following global reaction (Beltran et al., 1998; Guittoneau et al., 1990; Peyton and Glaze, 1988).$O_3 + H_2O \xrightarrow{hv} 2OH^{\bullet} + O_2$

It was reported that photolytic ozonation results in better degradation of some organic compounds (Leitus et al., 1990) than UV or ozonation alone. Ozone absorption of UV irradiation occurs primarily at 254 nm. Low pressure mercury vapor UV lamps are the most common sources of UV irradation used for this process. Many variables such as pH, temperature, scavengers in the influent, tubidity, UV intensity, lamp spectral characteristics and pollutant type(s) affect the effciency of the system.

v. ***UV radiation + Hydrogen Peroxide***: H_2O_2 molecule is photolyized by a radiation with a wavelength lower than 400 nm resulting in the cleavage of the molecule into hydroxyl radical with a quantum yield of two $OH^{\bullet}$ according to the following reaction (Legrini et al.,1983); $H_2O_2 \xrightarrow{hv} 2OH$

The medium pressure mercury vapor lamp are most commonly used lamps for this process because it has significant emittance within 200-250 nm, which is the primary absorption band for hydrogen peroxide. Some variables such as temperature, pH, concentration of H_2O_2, and presence of scavengers affect the production of hydroxyl radicals. For example, acid-base decomposition of hydrogen peroxide is undesirable because it consumes hydrogen peroxide without generating hydroxyl radicals. Another important issue is excessive amount of hydrogen peroxide which can act as a scavenger at relatively high concentrations. The negative effect of this reaction will depend on the residual concentration of hydrogen peroxide in the system.

vi. ***Ozone + UV radiation + Hydrogen Peroxide*** : This method is considered to be the most effective and powerful method which provides a fast and complete mineralization of pollutants (Mokrini et al., 1997).

vii. ***Fenton's Reagent*** : The reactivity of this system was first observed in 1894 by its inventor Fenton (Fenton, 1894). The process may be applied to industrial wastewaters containing a range of organic compounds such as phenols, formaldehyde, pesticides, wood preservatives, plastic additives, and rubber chemicals (Barbeni et al., 1987; Bishop et al., 1964, Eisenhauer, 1964; Flaherty and

Huang, 1994; Gau and Chang, 1996; Lipczynska et al., 195; Lipczynska, 1994; Lipczynska and Bolton, 1992; Lipczynska, 1991; Miller et al., 1996; Moza et al., 1988; Murphy et al., 1989), treatment plant sludge and contaminated soils with a reduction of toxicity, enhancement of biodegradability, and odor and color removal. Walling and Goosen (1973) have widely described the mechanism for the ferric ion catalyzed decomposition of hydrogen peroxide in acid solution. Hydroxyl radicals in this system is generated by means of the reaction between hydrogen peroxide and iron (II) salts as described in the following global reaction (Bigda 1995; Bossmans et al., 1998; Benitez et al., 1999);

$$H_2O_2 + Fe^{+2} \longrightarrow Fe^{+3} + OH^{\cdot} + OH^{-} \qquad k = 76.5\ l\ mol^{-1}\ s^{-1}$$

The rate of degradation of the organic pollutants by Fenton reaction could be increased when an irradiation source is present (Kiwi et al., 1993; Pulgarin and Kiwi, 1996) since photoreduction of Fe(III) ions to Fe(II), step (photo-Fenton) that produces new $OH^{\cdot}$ radicals occurs and regeneration of Fe(II) ions that can further react with more H_2O_2 molecules is possible (Perez et al., 2002). The photoreduction of Fe(III) follows the following equation;

$$Fe(OH)^{2+} + UV \longrightarrow Fe^{2+} + OH^{\cdot}$$

With $Fe(OH)^{2+}$ being the dominant Fe(III) species in solution at pH 2-3. High valence Fe intermediates formed through the absorption of visible light by the complex between Fe(II) and H_2O_2 are believed to enhance the reaction rate of oxidation production (Bossmann et al., 1998; Pignatello., 1992).

viii. ***Photocatalysis*** : This method is based on production of electron-hole pairs in the surface of the semiconductor (e.g.TiO_2 most commonly used semiconductor). These charged points react both with organic compounds and water. The destruction of organic compound is first done by the redox reactions and the generated hydroxyl radicals are in the latter effective (Gimenez et al., 1997). While direct O_3 or H_2O_2 photolysis need photons of short wavelength (< 310 nm), TiO_2 photocatalysis can take advantage of photons of wavelengths up to 380 nm (Hoffmann et al., 1995).

ix. ***Sonochemical Oxidation***: The effects of ultrasound waves on the hydrophilic chemical oxidations are due to the production of hydroxyl radicals during the cavitation-induced thermal decomposition of water (Shirgaonkar and Pandit, 1998). Most of the studies concerning ultrasound effects relate to the oxidation of aromatic compounds (Petrier et al., 1992) which is currently obtained with low rates. The combined action of sonochemistry and photochemistry has also been subject to researchers thereby, for example, Naffrechoux et al.(2000) observed an important enhancement of the degradation rate of phenol.

All reactions that do not result in the degradation of the pollutant are called "scavenging reactions". Mineral scavengers (bicarbonates, carbonates, phosphates etc.) present in many influents can drastically reduce the efficency of pollutant oxidation in direct proportion to their concentrations via following reactions (Duguet et al., 1989)

$$OH^{\bullet} + HCO_3^{-} \longrightarrow OH^{-} + HCO_3$$

$$OH^{\bullet} + CO_3^{2-} \longrightarrow OH^{-} + {}^{\bullet}CO_3$$

$$OH^{\bullet} + PO_4^{3-} \longrightarrow OH^{-} + {}^{\bullet}PO_4^{2-}$$

It has also been mentioned previously that reduced cations (e.g. iron) and excessive amounts of the primary oxidizers (e.g. ozone, hydrogen peroxide) can serve as significant scavengers of the hydroxyl radical (Hong et al., 1996; Zappi, 1995) which imply that an optimum dosage of oxidizers have to be used. Mass transfer limitations prevents formation of excessive amount of ozone, but risk of using excessive amount of hydrogen peroxide and thereby creating a scavenger is always possible (Hong et al., 1996; Zappi, 1995; Guittonneau et al., 1990).

Reactors and Light Sources

The adoption of AOPs based on UV systems requires the use of suitable UV sources and of appropriate photochemical reactors.

Reactors

Oxidation process is typically conducted in completely mixed tanks or plug flow reactors. The pollutant is introduced at one side and treated effluent exists at the other side. The oxidizing agent is either injected into the contaminated water just before it enters the tank or is dosed directly into the tank. The mixing of water with the oxidizing agent is provided either by mechanical agitation, pressure drop, or bubbling into the tank. Complete mixing is necessary to ensure full contact of the contaminants with the oxidizing agent for a minimum period of time.

Hydrogen peroxide and UV light are generally used together. Hydrogen peroxide is stored on site as a 35 to 50 %. The high solubility of H_2O_2 in water obviates the need for mixing equipment. The use of high-energy UV lamps, capable of imparting 500 W l^{-1} to the water, significantly reduces the size of the reactor used.

In case of ozone use (for example, O_3/UV system), a sparging device providing a good gas-liquid contact is necessary. For this purpose, ozone is added to the liquid as a gas, either through porous diffuser at the bottom of a tank, or through an injector where the pressure drop produced draws ozone into the injector and mixes it with the liquid.

Commercial AOP reactors developed by some companies for a full scale installation are available, for example Chemviron carbon and Wedeco. The reactor is normally provided with a device which allows a good mixing of dosed hydrogen peroxide. A careful attention must be paid to a proper sizing of the path length between quartz sleeve (housing the lamp) and the reactor walls so that more than 90% of emitted light must be absorbed by the solution. Therefore, the adoption of reflective reactor walls may be necessary. But, the cost and the potential fouling around reflectors reducing the benefits of reflectors cause controversy. Especially for the quartz sleeve, the problem of fouling is more relevant and it is essential that it remains clean during the operation. For this purpose, a quartz sleeve cleaning device should be provided.

Light Sources

During the photochemical oxidation processes, the light derived from the luminescence of excited atoms or molecules should be adsorbed by the reactant to achieve a photochemical reaction. The choice of effective light sources is dictated by the absorption spectrum of the reactant. The frequencies of emission of the source should correspond as closely as possible to the frequencies of absorption by the reactant. The different kinds of light sources produce emission lines or bands in various spectral regions from ultraviolet to visible and into the infrared; these regions may overlap from one source to another.

The light sources are distinguished by the means used to obtain excited states of the luminescent material: arc lamps, incandescent lamps, fluorescent lamps and tubes, lasers. In arc lamps, a gas is activated by an electric discharge between two electrodes: the atoms of the gas are excited by collision with the free electron of the arc (Noyes and Leighton, 1941). In incandescent lamps, a filament is heated to a high temperature by an electric current; the energy of excitation is furnished by the heat produced. In fluorescent lamps and tubes, an electric discharge in the gas phase provides the energy necessary to achieve of a fluorescent compound deposited on the inside wall of the tube (Elenbaas, ed., 1962). While the previously cited light sources emit a non-coherent light in all directions, the laser effect, the light amplification by stimulated emission of radiation laser, produces a beam of coherent light of both very strong intensity and well defined direction (Jones, 1969).

Power, which determines the rate of photons emitted and consequently, the rate at which the reactant is transformed chemically, is an another important criterion in the choice of a light source. Sources used most frequently in photo-oxidation processes are the arc lamps and fluorescent tubes, whose electric powers demand vary from 10 W to 60 000 W. Generally, the electric power of incandescent lamps is between 10 W and 1000 W (Braun et al., 1991).Optimum specific powers in the range 5-100 W l^{-1} are normally used in commercial reactors by means of UV lamps of 40 – 15 000 W (Bhattacharjee and Shah, 1998)

Thirdly, the geometry is another important quality in selecting of a light source. Various light sources are available which are either quasi-point sources, which may be used to produce a beam of variable cross-section, or are elongated (line, tube, coiled) sources which allow achievement of an appreciable surface of emission. A more detailed discussion on light sources is given by Braun et al. (1991).

Technology Limitations and Advantages

Some of the limitations associated with various oxidation processes are summarized below.

For example, in UV oxidation using H_2O_2, the process is only efficient at rather low wavelengths (below 200 nm). The aqueous stream being treated must provide good transmission of UV light, so that turbidity and high suspended solids concentration would not cause interferences. Scavengers and excessive dosages of chemical additives may inhibit the process. Heavy metal ions (higher than 10 mg l^{-1}), insoluble oil and grease, high alkalinity and carbonates may cause fouling of the UV quartz sleeves. Therefore, a good pretreatment of the aqueous stream should be necessary. Possible air emission problems with ozone (O_3) may be encountered in some UV oxidation systems.

At present, costs of AOPs are higher than competing technologies and depend on the degree of contaminant destruction required, flow rate, types and concentrations of contaminants (as they effect oxidizer selection, oxidizer dosage, UV light intensity etc.), requirements for pretreatment and/or post-treatment. The overall costs are represented by the sum of the capital costs, the operating costs (costs of reagents, electricity consumption, maintenance, labor, replacement of UV lamps etc.). The electricity cost, which is the most important cost component and based on electrical energy per order (EE/O) (Bolton et al., 1995), can be estimated using the following formula:

$$EE/O\ (kW.h/m^3) = (P \times t \times 1000) / [V \times 60 \times \log (C_{inf}/C_{eff})]$$

Where P is rated power (kW),V is the volume (in liters) of water treated, t is time (min), C_{inf} and C_{eff} are the initial and final concentrations of contaminant in COD (mg l^{-1}).

Some advantages include the following; $V/O_3/H_2O_2$ processes do not add to the pollutant load during treatment. On the other hand, conventional treatment alternatives such as chemical coagulation, carbon adsorption, biological treatment etc. merely transfer the waste from one medium to another leaving for example, excess biomass, contaminated chemical sludge, contaminated adsorbent residues for further disposal. Many refractive substances can be successfully treated using advanced oxidation processes.

A suitable application of AOP to wastewater treatment must consider that they make use of expensive reactants as H_2O_2, and/or O_3 and therefore it is obvious that the more economic biological treatment methods may still be the choice, whenever possible. The advantages offered by AOPs can be exploited to integrate biological treatments by an oxidative degradation of toxic or refractory substances entering or leaving the biological stage.

Case Studies

Decolourization of Textile Effluents

Coloured effluent in the dye manufacturing and dye consuming industries is a result of ineffective batch operations. Approximately 2% of dyes that are produced and 10% during the textile colourization process are eventually lost resulting in coloured aqueous effluent (Easton, 1995). The magnitude of the problem is obviously high considering that over $7x10^5$ tones of approximately 10,000 types of dyes and pigments are produced annually worldwide (Robinson et al., 2001) and synthetic dyes are persistent in the environment due to their recalcitrant and toxic nature (McKay, 1979; Pagga and Brown, 1986). Textile effluents are also problematic from the aesthetical point of view because the presence of very low concentrations of dyes in effluent is highly visible and undesirable (Nigam et al., 2000).

The colour and chemical composition of textile wastewater is usually subject to daily and seasonal variations that are dictated by production routines and by fashion cycles. The amount of dye loss due to the incomplete exhaustion of dyes on to the fiber depends on dyestuff type (cationic, nonionic, mordant, acid, basic, disperse, direct, vat, sulfur, azoic, pigment and reactive dyes) application route and the depth of shade required (Willmott, 1997). Acid dyes are water-soluble anionic dyes with chromophore (color-bearing) groups substituted with acidic functional groups such as nitro-, carboxyl-, and sulfonic acid. Basic dyes are cationic types with chromophores typically having amino groups. Direct dyes are highly water-soluble salts of sulfonic acid of azo dyes. Reactive dyes, similar to direct dyes, are also highly water-soluble anionic dyes, but have excellent wet fastness with binding to the textile fibers via covalent bonds. Disperse dyes are substantially water-insoluble nonionic dyes for the application to the hydrophobic fibers from aqueous dispersion. Azo dyes, characterized by nitrogen to nitrogen double bonds, account for up to 70% of all textile dyestuffs produced, and are the most common chromophore of the reactive dyes (Carliell et al., 1995). An extensive review regarding the properties of dyestuff are given by Hao et al.(2000).

Typically, soluble coloured compounds are responsible for 24 to 35 % of the waste COD and 90 to 95 % of the waste colour (total absorbance); the suspended and colloidal coloured compounds account for the remaining small fraction of the color (Tzitzi et al., 1994). The complex dye solution is reflected by the light absorption in a wide UV-VIS spectrum range from 250 to 600 nm (Matsui et al., 1981). Furthermore, the intensity of the color varies with pH (Zhang et al., 1999).

There are two possible approach that can be applied to combat with dyestuff contaminated effluent; i) use of end of pipeline technologies meaning that developing dedicated on-site or in-plant facilities to provide partial or full treatment of the coloured effluent before discharge into either receiving media or common sewage channel ii) pollution prevention (2P) technologies that allow partial or full re-use of the coloured water. The recent advances of "Cleaner Technology" may assist the dye waste minimization and use of "clean" technologies (Hickman, 1993).

Currently, textile effluent treatment involves physical and/or chemical processes (coagulation/flocculation, membrane technologies etc.) followed by a biological process which result in accumulation of concentrated chemical sludge that are very costly to dispose. The combination of physical-chemical and biological methods (so called integrated processes) have limited success (Azbar et al., 2004; Balcioglu and Arslan, 1997; Wu et al., 1998). Textile wastewaters exhibit low BOD to COD ratios (< 0.1) indicating non-biodegradable nature of dyes and Wilmott et al.(1998) have claimed that aerobic biological degradation is not always effective for textile dye contaminated effluent. Furthermore, the reductive cleavage of the azo bond under anaerobic conditions result in aromatic amines which are more toxic than intact dye molecules (Chung et al., 1978; Weber and Wolfe, 1987). Therefore, the treatment of synthetic dyestuff cannot depend on biodegradation alone. Traditional physical-chemical techniques such as coagulation/flocculation, ultra filtration, reverse osmosis, ion exchange and adsorption on various adsorbents (activated carbon, peat, fly-ash and coal, wood chips, silica gel, corncob, barley etc.) have efficiently been used for the purpose of color removal from textile effluents. Coagulation may not remove highly soluble dyes, for example, while acid, direct, vat, mordant and reactive dye usually coagulate, but cationic dyes do not coagulate at all, and it may be good for disperse dyes with the production of a large quantity of chemical sludge. In addition, for lime coagulation, some toxic byproducts may be formed due to a drastic increase in pH, for example, the formation of aniline and chloroanilines (Sarasa et al., 1998). Activated carbon is the most commonly applied method for color removal especially for cationic, mordant, and acid dyes and a slightly lesser extent, dispersed, direct, vat, pigment and reactive dyes (Raghavacharya, 1997). Carbon adsorption of dyes is only successful in some classes of dye (ionic type and hydrophobicity) for a given type of carbon in a certain pH range (Hao et al., 2000).

Nevertheless, these techniques are non-destructive, since they just transfer the pollutant from water to solid matrix.

Other emerging techniques, such as advanced oxidation processes (AOPs) may have potential for decolourization. Although AOPs have a significant advantages over conventional treatment methods since chemical oxidation do not result in high amount of either chemical or biological sludge and almost complete demineralization of organics is possible, one of the main handicap with AOPs presently lies in the high cost of reagents or energy sources like ultraviolet light (Azbar et al., 2004; Galindo et al., 2001).

AOPs have been thoroughly and comparatively evaluated for a variety of organic compounds and wastewaters in the past but with several limitations. Also many investigations have demonstrated that AOPs are effectively removing color and partially organic content of dyestuffs (Clark and Knowles, 1982; Bolton et al., 1995; Benitez et al., 1997; Legrini et al., 1983; Marechal et al., 1997). Some reactive, basic, and direct dyes are known to be easily subject to UV/H_2O_2 oxidation at pH 5 with more than 95% of colour removal, whereas disperse and vat dyes are only partially decolorized (Yang et al., 1998). Ozonation alone is not effective for disperse dyes (Solozhenko et al., 1995). A combination of these processes is necessary to achieve the desirable goal. For example, the degree of the inhibition for a synthetic dye in a biological-activated sludge system has been reduced from 47% to 10% after a pretreatment with O_3/UV or $O_3/H_2O_2/UV$ (Ledakowicz and Gonera, 1999).

Azbar et al. (2004) carried out a comparative study on color and COD removal from acetate and fiber dyeing effluent. In this study, various advanced oxidation processes (O_3, O_3/UV, H_2O_2/UV, $O_3/H_2O_2/UV$, Fe^{+2}/H_2O_2) and chemical treatment methods using $Al_2(SO_4)_3.18H_2O$ $FeCl_3$ and $FeSO_4$ for the Chemical Oxygen Demand (COD) and color removal from a polyester and acetate fiber dyeing effluent is undertaken. Advanced Oxidation Processes (AOPs) showed a superior performance compared to conventional chemical treatment, which maximum achievable color and COD removal for the textile effluent used in this study was 50% and 60 %, respectively. Although $O_3/H_2O_2/UV$ combination among other AOPs methods studied in this paper was found to give the best result (99% removal for COD and 96% removal for color), use of Fe^{+2}/H_2O_2 seems to show a satisfactory COD and color removal performance and to be economically more viable choice for the acetate and polyester fiber dyeing effluent on the basis of 90% removal. Operating cost is a main issue for AOPs and relatively high, for example, Azbar et al. (2004) calculated the operating costs per m^3 (excluding UV lamp and labor) as follows; \$0.23 for Fenton's reagent, \$1.26 for H_2O_2/UV, \$6.38 for O_3/UV and \$6.54 for $O_3/H_2O_2/UV$ processes that all provided over 90% COD removal. More or less, the literature reports regarding cost of AOPs are in the same parallel varying between \$3.2 and \$7.7 per m^3.

Decolourization of textile effluents is of major environmental concern that remains to be solved.

Detoxicifation of Olive Mill Effluent (OME)

Ironically, while the olive oil itself provides health during its consumption, its resulting by-products (olive mill effluent, OME) during olive processing pose a serious environmental pollution especially in the Mediterranean that accounts for approximately 95 % of worldwide olive oil production. The olive mill waste products (OMW) derived from olive oil extraction are an aqueous effluent (vegetation water) and a solid residue, mainly containing the olive skin and stone (olive husk) composed of many complex coloured and recalcitrant compounds (aromatics: 1-10 g m^{-3}) with high organic loads (Chemical Oxygen Demand, COD: 100-300 g m^{-3}) (Andreoni et al., 1986; Boari et al., 1984; Gonzalez et al., 1990; Tuncel and Nergiz, 1993). Conventional press extraction process results in 0.5 - 0.8 m^3 per ton of olives, whereas the waste stream volume is 1.18 - 1.68 m^3 per ton of olives from 3-phase continuous centrifugal process. Production facilities and wastes are usually associated with strongly odorous volatile compound emissions. Olive mill wastes are a significant source of environmental pollution especially in some important olive oil producing countries such as Spain, Italy, Greece, Tunisia, Turkey, Lebanon, Syria and Portugal (Azbar et al., 2002; Beccari et al., 1996). When disposed into the environment, OME create serious deteriorations such as colouring of natural waters, a serious threat to the aquatic life, pollution in surface and ground waters, alterations in soil quality, phytoxicity and odor nuisance.

OME is tough to deal with and biological treatment of OME is a challenge because of high solids concentration in the effluent and biologically resistant compounds such as tannins, polyphenols, polyalcohols, pectins and lipids in OME (Hamdi, 1996; Zouari, 1998). Many central municipal wastewater treatment plants in Turkey had to be shut down during the olive campaign period due to discharges of untreated OME with disastrously high organic and recalcitrant content into the sewage (Azbar et al., 2002). Several methods were tested for the treatment of OME such as mechanical, physical, chemical, biological and thermal methods. An advisable single treatment method that is economically viable has not been suggested yet. Evaporation in storage lagoons is currently the common disposal method applied in most countries as long as sufficiently large land areas are available.

Benitez et al. (1999) studied the degradation of the pollutant organic matter present in OME (using 10 times diluted sample). This study shows that ozonation of OME achieved a moderate reduction in the COD, and significant removal of aromatics and total phenolics. Rivas et al. (2000) reported that application of pre-ozonation (45 mg l^{-1}) to OME sample (diluted 1:25 with synthetic urban wastewater) for 35 min before aerobic degradation achieved a better biodegradability of OME. Benitez et al. (2001) applied a set of oxidation processes (single UV, UV/H_2O_2, Fe^{+2}/H_2O_2, Fe^{+2}/H_2O_2/UV) for the purpose of treating black olive oil washing waters. While single UV application resulted in only 35% removal of organic matter, hydroxyl radical enhanced this reduction with values in the range 41-76%. Similarly, removal of aromatics at a level of 20% was increased up to 52% via use of combined UV+H_2O_2. Rivas et al. (2001) used wet air oxidation for OME (diluted 1:10) at high temperatures and pressures at the original pH of the effluent (pH=5.3). They reported that wet air oxidation is a feasible process and OME becomes more amenable for further biodegradation via wet air oxidation and use of hydroxyl radicals promoters such as hydrogen peroxide resulted in a significant enhancement of the process. In addition, use of two commercially available catalysts (platinum supported on alumina and copper oxide supported on active carbon) showed not only an improvement in the organic removal but also a high degree of the mineralization of the influent.

Kestioglu et al. (2004) studied the efficiency of alternative pretreatment methods for an olive mill effluent (OME) in producing a final effluent conforming to Turkish Water Pollution Standards for discharge into the public sewage. For this purpose, various treatment alternatives such as acid cracking, chemical coagulation, adsorption and Advanced Oxidation Processes (AOPs) were studied. A significant amount of Chemical Oxygen Demand (COD), total phenol and oil&grease removal were obtained by acid cracking followed by chemical coagulation either using alum or iron chloride salts. Among the alternative chemicals, both Al_2SO_4 and $FeCl_3$ resulted in the same COD and total phenol removal efficiency in this study (94-95% COD removal and 90-91% total phenol removals). Both H_2O_2/UV and O_3/UV combinations studied in this paper were found to give practically same COD and total phenol removal efficiencies (over a 99 % removal for both COD and total phenol).

Treatment of Domestic Wastewater

Biological wastewater treatment is currently the most widely used method for removing organics and nutrients from domestic effluents. Among the alternatives treatment flow schemes, A^2/O activated sludge system where a combination of an anaerobic, an anoxic and an aerobic environment is used to maintain biological nitrification of excess ammonia, denitrification of produced nitrates and excess phosphorus uptake by phosphate-accumulating organisms (PAO). Biological methods especially advanced nutrient removal systems such as A^2/O are very complicated and difficult to stabilize and optimize especially when a toxic component is involved in the influent. In addition, these processes require a larger treatment system and longer time than alternative oxidation processes. In addition to industrial treatment utilizing powerful hydroxyl radicals, treatment of domestic wastes and/or sludge may be feasible.

Although the biological wastewater treatment processes are able to remove more than 90% of the BOD (biological oxygen demand) in wastewater, there are some persistent organic compounds that are not easily removed by biological treatment and need further purification (Schroder, 1998). Wang et al. (2001) used UV/H_2O_2 to evaluate its suitability to eliminate the organic pollutants in the non-chlorinated effluents of a secondary wastewater treatment plant. The authors applied filtration before photodegradation experiments. They reported the best oxidation efficiency was obtained under acidic conditions (pH 5) and the optimum H_2O_2 dose was between 0.01% and 0.1% for the oxidation process.

In recent years, electrochemical treatment has gained increasing interest. This process can completely convert organic pollutants into gases such as N_2, CO_2, etc. The most of the electrochemical applications were on a laboratory scale except for a few industrial applications. Vlyssides et al. (2002) applied an electrochemical treatment to domestic wastewater using 0.8% (w/v) sodium chloride as electrolyte and an electrolytic cell with Ti/Pt as anode and Stainless steel 304 as cathode. Due to the strong oxidizing potential of the chemicals produced (chlorine, oxygen, hydroxyl radical and other oxidants), the organic pollutants and nutrients (organic nitrogen, phosphorus) were wet oxidized to carbon dioxide (T= 40 °C and pH = 9), and nitrogen as well as phosphorus was precipitated as $Ca_3(PO_4)_2$. COD was reduced by 89%, volatile suspended solids (VSS) by 90%, ammonia nitrogen by 82% and total phosphorus by 98%. The efficiency of electrolysis went up to 35 g COD_t/(h m^2 A) and the energy consumption to 12.4 kWh/kgCOD_t .

Feng et al. (2003) tested the electrochemical treatment method for the purpose of treating domestic wastewater. The authors applied 3 mA cm^{-2} current density with 500 v pulse voltage at a frequency of 25 kHz using a plot reactor with a 0.3 m^3 h^{-1} capacity. The concentrations of T-N, NH_4-N, T-P, and COD in domestic waste after treatment by the electrochemical treatment system were reduced by 73, 81, 99 and 86%, respectively.

The efficiency of TiO_2, photofenton and the modified photo-fenton reagent in the presence of solar irradiation for the purpose of organic reduction from a synthetic municipal wastewater was evaluated by Kositzi et al. (2004). The authors studied the effect of the type of the catalyst, as well as the influence of oxidizing agents such as H_2O_2 and $Na_2S_2O_8$. In this study, the accumulation energy of of 50 kJL-1 with the synergetic effect of 0.2 gL^{-1} TiO_2 P-25 with H_2O_2 and $Na_2S_2O_8$ provided a 55% and 73% reduction of the initial organic carbon content, respectively. The authors concluded that the photo-fenton process appeared to be more efficient for this type of wastewater in comparison to the TiO_2/oxidant system. An accumulation energy of 20 kJL^{-1} provided an 80% reduction of the organic content and the presence of oxalate in the Fe^{3+}/H_2O_2 system resulted in an additional improvement of the photocatalytic efficiency.

The authors of this study applied UV/H_2O_2 process to treat domestic wastewater in their laboratory (unpublished data). A hydrogen peroxide dose of 50 mg l^{-1} were applied to the domestic wastewater (COD_{inf} = 500 mg l^{-1}) at pH 3 under laboratory conditions. A set of UV lamps varying from 8W up to 48 W in a 2.8 L photo reactor were used during the 60 min of reaction time (batch process). It was observed that as the power of UV lamp increased, the treatment efficiency increased and the reaction time decreased. It was possible to decrease the initial COD of 500 mg l^{-1} down less than 50 mg l^{-1} in about 50 min reaction time. A second order reaction kinetic was found to fit better according to the data observed during this study. In the second part of the study, varying initial COD_{inf} values (500, 750, 1000, 1250, 1500 and 2000 mg l^{-1}) were used under constant UV lamp power (32 W) to find the UV dose needed to eliminate each mg of COD removed. COD removal and UV dose relation was found as 60 mg COD per UV dose (W).

CONCLUSIONS

- Advanced oxidation processes represent a powerful mean for the abatement of refractory and/or toxic pollutants in wastewaters.
- Advanced oxidation processes although making use of different reacting systems, are all characterized by the same chemical feature: production of $OH^{\bullet}$ radicals, which are extraordinarily reactive species attacking the most part of organic molecules with rate constants usually in the order of $10^6 - 10^9$ M^{-1} s^{-1}.
- AOPs offer different possible ways for $OH^{\bullet}$ radicals production thus allowing a better compliance with the specific treatment requirements.
- pH of the solution influences the rate of radical generation for the system O_3/H_2O_2, H_2O_2/UV and O_3/UV, proper value of this parameter must therefore be chosen.
- The presence of scanvengers like carbonates can result in a significant reduction of the efficiency.
- One should consider the higher operating and capital costs therefore combined use of AOPs with biological methods should be first choice, when possible.
- Mass transfer limitations during, for example, O_3/H_2O_2 and O_3/UV systems should be taken into account during the design.

- Referring to the pollution load of wastes normally expressed as COD, only wastes with relatively small COD contents (≤ 5 g l^{-1}) may be suitably treated by means of these techniques as the COD content increases, the consumption of expensive reactants significantly increase making the process economically unviable.
- The identification of reaction intermediates, development of rate expressions, identification of scale-up parameters, the improvement of these techniques for a more efficient exploitation of solar radiation and criteria for cost effectiveness and maximum destruction efficiency should be in the agenda of researchers in the future.

REFERENCES

Adams, C.D., Scanlan, P.A., Secrist, N.S. (1994). Oxidation and biodegradability enhancement of 1,4-dioxane using hydrogen peroxide and ozone. *Environ. Sci. Technol.*, 28: 1812-1818.

Andreoni, V., Ferrai, A., Ranaldi, G. and Sorlini, C. (1986). The influence of some phenolic acids present in oil wastes on microbic groups for methanogenesis. *In Simposio International Sobre Valoriazcion de los Subproductos de Olivar: Alpechin y Orujo*, 5-6-7 March, Sevilla, Spain.

Andreozzi, R., Caprio, V., D'Amore, M.G., Insola,A. (1992). The kinetic of Mn(II) catalysed ozonation of oxalic acid in aqueous solution.*Water Res.*, 26(7):917-921.

Andreozzi, R., Caprio, V., D'Amore, M.G., Insola,A. (1995). Manganese catalysis in water pollutants abatement by ozone. *Environ.Technol.*, 16:885

Azbar, N., Cevilan,I., and Turan, I.,2002. Effect of OOMW on central treatment plant and evaluation of the pretreatment requirements. 1st International Workshop on Environmental Problems in Olive Oil Production and Solutions, Azbar, N, Vardar, N, Akın, M. and Cevilan, I. (Eds)., 7-8-9 June, 2002, *Edremit, Balikesir,Turkey, Proceedings,* p.1-6.

Azbar, N., Yonar, T., Kestioglu, K. (2004). Comparison of various advanced oxidation processes and chemical treatment methods for COD and color removal from a polyester and acetate fiber dyeing effluent. *Chemosphere*, 55(1):35-43.

Barbeni, M., Minero, C., Pelizzetti, E., Borgaerllo, N., Serpone, N. (1987). Chemical degradation of chlorphenols with Fenton's reagent. *Chemosphere.*, 16:2225-2237.

Balcioglu, I.A., Arslan, I. (1997). Treatment of textile wastewater by heterogenous photocatalytic oxidation processes. *Environ.Technol.*,18:1053.

Beccari, M., Bonemazzi, E., Majone, M. and Riccardi, C. Interactions between acidogenesis and methanogenesis in the anaerobic treatment of olive mill effluents. *Water Res.* 1996;30: 183-189.

Beltran, F.J., Ovejero, G., Acedo, B. (1993). Oxidation of aztrazine in water by UV radiation combined with H_2O_2. *Water Res.*, 27:1013-1021.

Beltran, F.J., Encinar, J.M., Alonso, M.A. (1998). Nitroaromatic hydrocarbon ozonation in water. 2. Combined ozonation with hydrogen peroxide or UV radiation. *Ind. Eng. Chem. Res.*, 37:32-40.

Benitez, F.J., Heradia, J.B., Acero, J.L., Pinila, M.L. (1997). Simultaneous photodegradation and ozonation plus UV radiation of phenolic acids-major pollutants in agro-industrial wastewaters. *J. Chem. Tech. Biotech.*, 253-260.

Benitez, F.J., Beltran-Heredia, J., Acero, J.L., Rubio, F.J. (1999): Chemical decomposition of 2,4,6-trichlorophenol by ozone, Fenton's reagent and UV radiation. *Ind. Eng. Chem. Res.*, 38(4):1341-1349.

Benitez, F.J., Beltran-Heredia, J., Torregrosa, J., Acero, J.L. (1999). Tretament of olive mill wastewaters by ozonation, aerobic degradation and the combination of both treatments. *Journal of Chemical Technology and Biotechnology.*, 74:639-646.

Benitez, F.J., Acero, J.L., Gonzalez, T., Garcşia, J. (2001). Organic removal from wastewaters of the black olive industry by chemical and biological procedures., *Process Biochemistry*,37:257-265.

Bhattacharjee, S. and Shah, Y.T. (1998). Mechanism for advanced photooxidation of aqueous organic waste compounds. *Rev. Chem. Eng.*, 14(1):1.

Bigda, R.J. (1995). Consider Fenton's chemistry for wastewater treatment. *Chem. Eng. Prog.*, 91(12):62-66.

Bishop, D.F., Stern, G., Fleishman, M., Marshall, L.S. (1964). Hydrogen peroxide catalytic oxidation of refractory organics in municipal wastewaters. *Am. Chem. Soc., Div. Wastewater Chem.*, Preprints 4 (2):85-90.

Boari, G., Brunetti, A., Passino, R. and Rozzi, A. Anaerobic digestion of olive oil mill wastewaters. *Agricultural Wastes*, 1984;10:.93-110.

Bolton, J.R., Cater, S.R. (1994). In Surface and Aquatic Environmental Photochemistry. G.Helz, R.Zepp, D.Crosby (Eds.),Ch.33, pp.467-490,*CRC Press*, Boca Raton.

Bolton, J.R., Bircher, K.G., Tumas, W., Tolman, C.A., 1995. Figures of merit for the technical development and application of advanced oxidation processes. *J. of Adv. Oxidation Technol.* 1-11.

Bossmann, S.H., Oliveros, E., Göb, S., Siegwart, S., Dahlen, E.P., Payawan, Jr L., Straub, M., Wörner, M., Braun, A.M. (1998). New evidence against hydroxyl radicals as reactive intermediates in the thermal and photochemically enhanced Fenton reactions. *J. Physc.Chem.* A., 102(28):5542-5550.

Braun,A.M., Maurette, M.T., Oliveros, E. (1991). *Photochemical Technology*. John Wiley & Sons. New York.

Carliell, C.M., Barclay, S.J., Naidoo, N., Buckley, C.A., Mulholland, D.A., Senior, E. (1995). *Microbial decolorization of a reactive azo dye under anaerobicconditions.* Warter SA, 21(1):61-69.

Chung, K.T., Fulk, G.E., Egan,M. (1978). Reduction of azo dyes by intestinal anaerobes. *Appl. Environ. Microbiol.* 35, 558-562.

Clarke, N., Knowles, G. (1982). High purity water using H_2O_2 and UV radiation. *Effluent and Water Treatment Journal.* 335-341.

Crittenden J.C., Hu S, Hand, D.W. (1999). Advanced oxidation model for H_2O_2/UV process in a completely mixed batch recator.*Water Res.*, 33 (10), 2315-2328.

Duguet, J.P., Anselme, C., Mallevialle, J. (1989). New advances in oxidation processes :Some examples of application of ozone/hydrogen peroxide combination for the removal of micropollutants for drinking water. *Water Nagoya, ASPAC IWSA.*

Easton, J. (1995). The dye maker's view. In: Cooper, P.(Ed). Colour in dyehouse effluent. Bradford, UK:*Society of Dyers and Colourists*, p.11.

Eisenhauer, H.R. (1964). Oxidation of phenolic wastes., J.Water Pollut.Cont.Fed., 36:1117-1127. Fluorescent Lamps and Lighting, W.Elenbaas, ed., *Hume Press*, London, 1962.

Esplugas, S., Gimenez, J., Contreras, S., Pascual, E., Rodrigues, M. (2002). Comparison of different advanced oxidation processes for phenol dagradation. *Water Res*., 36;1034-1042.

Feng, C., Sugiura, N., Shimada, S., Maekawa, T. (2003). Development of a high performance electrochemical wastewater treatment system. *Journal of Hazardous Materials* B103., 65-78.

Fenton, H.J.H. (1894). Oxidation of tartaric acid in presence of iron. *J. Chem.Soc*., 65:899-910.

Flaherty, K.A., Huang, C.P. (1994). Batch and continuous flow applications of Fenton's reagent and Fenton-like chemistry for the treatment of refractory textile wastewaters., *Chem.Oxid.,* 3:112-133.

Galindo, C., Jacques, P., Kalt, A., 2001. Photochemical and photocatalytic degradation of an indigolid dye: A case study of acid blue 74 (AB74). *Journal of Photochemistry and Photobiology A: Chemistry*. 141, 47-56.

Gau, S.H., Chang, F.S. (1996). Improved Fentonmethod to remove recalcitrant organics in landfill leachate., *Water Sci. Technol*., 34(7-8):455-462.

Gimenez, J., Curco, D., Marco, P. (1997). Reactor modelling in the photocatalytic oxidation of wastewater. *Water Sci. Tech*., 35(4):207-213.

Glaze W.H., Kang, J.W. (1989). Advanced oxidation processes. Description of a kinetic model fro the oxidation hazardous materials in aqueous media with ozone and hydrogen peroxide in a semibatch reactor. *Ind. Eng. Chem. Res*., 28:1578-1580.

Gonzalez, M.D., Moreno, E., Quevedo, J. and Ramos, A. Studies on antibacterial activity of waste waters from olive oil mills (alpechin): inhibitory activity of phenolic and fatty acids. *Chemosphere* 1990; 20: 423-432.

Guittoneau, S., De Laat, j., Duguet, J.P., Bonnel, C., Dore M. (1990). Oxidation of parachloronitrobenzene in dilute aqueous solution by O_3+UV and H_2O_2+UV: a comparative study. *Ozone Sci. Eng*., 12 :73-94.

Hamdi, M. (1996). Anaerobic digestion of olive oil wastewaters. *Process Biochemistry*. 31 (2):105-110.

Hao, O.J., Kim, H., Chiang P.C., 2000. Decolorization of wastewater. *Critical Reviews in Environ. Sci. Technol*. 30(4), 449-505.

Hickman, W.S. (1993). Environmental aspects of textile processing. *J.Soc.Dyers Colourists*. 109(1):32-37.

Hoigne, J. (1997). Inter-calibration of OH radical sources and water quality parameters. *Water Sci.Technol*., 35(4), 1-8.

Hoigne, J. (1998). Chemistry of aqueous ozone and transformation of pollutants by ozone and advanced oxidation processes, in:J.Hrubec (Ed.), *The Handbook of Environmental Chemistry, vol.5, part C, Quality and Treatment of Drinking Water, Part II, Springer, Berlin Heidelberg*.

Hoigne, J., and Bader, H. (1983). Rate constants of reaction of ozone with organic and inorganic compounds in water. Part II. Dissociating organic compounds. *Water Res*., 17;185-194.

Hoffmann, M.R., Martin, S.t., Choi, W., Bahnemann, D.W. (1995).Environmental application of semiconductor photocatalysis. *Chem.Rev*., 95:69-96.

Holf, C., Sigl, G., Specht , O., Wurdack, I., Wabner, D. (1999). Oxidative degradation of AOX and COD by different advanced oxidation processes:A comparative study with two samples of pharmaceutical wastewater. *Water Sci.Technol.*, 35:257-264

Hong, A., Zappi, M.E., Kuo, C.H., Hill, D.O. (1996). Modeling the kinetics of illuminated and dark advanced oxidation processes. *ASCE J. Environ.Eng.*, 122(1) : 58-62.

Jones, W.J. (1969). *Quarterly Rev.*, 23: 73.

Kestioglu, K., Yonar, T., azbar, N. (2004). Feasibility of Physico-Chemical Treatment and Advanced Oxidation Processes (AOPs) as a Means of Pretreatment of Olive Mill Effluent (OME), *Process Biochemistry* (In Press).

Kiwi, J., Pulgarin, C., Peringer, P., Gratzel, M. (1993). Beneficial effect of homogenous photo-Fenton pretreatment upon the biodegradation of anthraquinone sulfonate in wastewater treatment. *Appl. Catal.B:Environ.*, 3:85-99.

Kositzi, M., Poulios, I., Malato, S., Caceres, J., Campos, A. (2004). Solar photocatalytic treatment of synthetic municipal wastewater. *Water Res.*, 38:1147-1154.

Ku, Y., Wang, L., Shen, V. (1998). Decomposition of EDTA in aqeous solution by UV/H_2O_2 process. *J.Hazard.Mater.*, 60:41-45.

Ledakowicz, S., Gonera, M. (1999). Optimisation of oxidant dose for combined chemical and biological treatment of textile wastewater. *Water Res.* 33(1), 2511-2516.

Legrini, O., Oliveros, E., Braun, A.M. (1983). Photochemical processes for water treatment. *Chem. Rev.* 93, 671-698.

Leitus, E., Zeff, J.D., Smith, M.M., Crosby, D.G. (1990). An investigation into the chemistry of the UV-ozone purification process, Final Report to the National Science Foundation on Grant ENV 76-24652, *US Government Printing Office*, Washington, DC.

Lipczynska-Kochany, E., Sprah, G., Harms, S. (1995). Influence of somegroundwater andsurface waters constituents on the degradation of 4-chlorphenol by the Fenton reaction., *Chemosphere.*, 30(1): 9-20.

Lipczynska-Kochany, E. (1994). Degradation of aromatic pollutants by means of the advanced oxidation processes in a homogeneous phase: photoloysis in the presence of hydrogen peroxide versus the Fenton reaction., *Chem. Oxid.*, 3:12-27.

Langlais, B., Reckhow, D.A., Brink, D.R. (1991). Ozone in Water Treatment., Application and Engineering. *Lewis Publishers*, Chelsea, MI.

Legrini, O., Oliveros, E., Braun, A.M. (1983). Photochemical processes for water treatment. *Chem. Rev.*, 93:671-698.

Lipczynska-Kochany, E. (1991). Degradation of aqueous nitrophenols and nitrobenzene by means of Fenton reaction., *Chemosphere.*, 22(5-6):529-536.

Marechal, M.L., Slokar, Y.M., Taufer, T., 1997. Decolorization of chlorotriazine reactive azo dyes with H_2O_2/UV. *Dyes And Pigments*. 33:281

Moza, P.N., Fytianos, K., Samanidou, V., Korte, F. (1988). Photodecomposition of chlorophenols in aqueous medium in thepresence of hydrogen peroxide. Bull. Environ. *Contam. Toxicol.*, 41(5):678-682.

Mokrini, A., Ouose, D., Esplugas, S. (1997). Oxidation of aromatic compounds with UV radiation/ozone/hydrogen peroxide. *Water Sci.technol.*, 35:95-102

Murphy, A.P., Boegli, M.K., Price, M.K., Moody, C.D. (1989). A Fenton-like reaction to neutralize formaldehyde waste solutions. *Ozone:Sci. Eng.*, 15:166-169.

Matsui, M., Kobayashi, K., Shibata, K., Takase, Y. (1981). Ozonation of dyes.II. Ozone treatment of 4-phenylazo-l-naphthol. *J.Soc.Dyers Colourists*. 97(5):210-213.

McKay, G. (1979). Waste colour removal from textile effluents. *J. Am. Dyes. Rep.*68, 29-36.

Mokrini, A., Oussi, D., Esplugas, S. (1997). Oxidation of aromatic compounds with UV radiation/ozone/hydrogen peroxide. *Water Sci. technol.*, 35(4):95-102.

Naffrechoux, E., Chanoux, S., Petrier, C., Suptil, J. (2000). Sonochemical and photchemical oxidation of organic matter. *Ultrasonics Sonochemistry.*, 7:255-259.

Nigam, P., Armour, G., Banat, I.M., Singh, D., Marchant, R. (2000). Physical removal of textile dyes and solid state fermentation of dye-adsorbed agricultural residues. *Bioresource Technol.* 72, 219-226.

Noyes, Jr. W.A. and Peighton, P.A. (1941). The Photochemistry of Gases. Dover, New York.

Pagga, U., Brown, D. (1986). The degradation of dyestuffs part II: Behaviour of dyestuffs in aerobic biodegradation tests. *Chemosphere.* 15(4), 479-491.

Pera-Titus, M., Garcia-Molina, V., Banos, M.A., Gimenez, J., Esplugas, S. (2004). Degradation of chlorophenols by means of advanced oxidation processes : a general review. *Apllied Catalysis.*, 47: 219-256.

Perez, M., Torrades, F., Garcia-Hortal, J.A., Domenech, X., Peral, J. (2002). Removal of organic contaminants in paper pulp treatment effluents under Fenton and photo-Fenton conditions. *Applied Catalysis B: Environmental*, 36:63-74.

Petrier, C., Micolle, M., Merlin, G., Luche, J-L., rverdy, G. (1992). Charateristics of pentachlorophenate degradation in aqueous solutions by means of ultrasound. *Environ.Sci.Technol.*,26 :1639.

Peyton, G.R., Glaze, W.H. (1988). Destruction of pollutants in water with ozone in combination with ultraviolet radiation. Part 3. Photolysis of aqueous ozone. Mechanism of photolytic ozonation. *Environ.Sci.Technol.*, 22 :761.

Pignatello, J.J. (1992). Dark and photassisted Fe^{+3}-catalyzed degradation of chlorophenoxy herbicides by hydrogen peroxide. *Environ.Sci.Technol.*, 26 :944-951.

Pulgarin, C., Kiwi, J. (1996). Overview on photocatalytic and electrocatalytic pretreatment of indsutrial non-biodegradable pollutants and pesticides. *Chimia,* 50:50.

Raghavacharya, C. (1997). Colour removal from industrial effluents-A comparative review of available technologies. *Chem. Eng. World.* 32, 53-54.

Rivas, F., Beltran, F.J., Gimeno, O. (2001). Joint treatment of wastewater from table olive processing and urban wastewater. Integrated ozonation-aerobic oxidation., *Chem.Eng.Technol.*, 23(2):177-181.

Rivas, F., Beltran, F.J., Gimeno, O., Acedo, B. (2001). Wet air oxidation of wastewater from olive oil mills.*Chem.Eng.Technol.*, 24(4): 415-421.

Sarasa, J., Rıche, M.P., Ormad, M.P., Gimeno, H., Puig, A., Ovelleiro, J.L. (1998). Treatment of a wastewater resulting from dyes manufacturing with ozone and chemical coagulation. *Wat.Res.*, 32(9):2721-2727.

Schroder, H. (1998). Characterization and monitoring of persistent toxic organics in the aquatic environment. *Water Res.* 38:151-158.

Shirgaonkar, I.Z., Pandit, A.B. (1998). Sonophotochemical destruction of aqueous solution of 2,4,6-trichlorophenol. *Ultrason.Sonochem.*, 5 :53.

Solozhenko, E.G., Soboleva, N.M., Goncharuk, V.V. (1995). Decolourization of azodye solutions by Fenton's oxidation. *Water Res.* 29(9).2206-2210.

Stachlein, S., Hoigne, J. (1982). Decomposition of ozone in water. Rate of initiation by hydroxide ions and hydrogen peroxide. *Environ. Sci. Technol.*, 16 :676-681.

Robinson, T., McMullan, G., Marchant, R., Nigam, P. (2001). Remediation of dyes in textile effluent : a critical review on current treatment technologies with a proposed alternative. *Bioresource Technology*, 77:247-255.

Tuncel, G. and Nergiz, C.,1993. Antimicrobial effect of some olive phenols in a laboratory medium. *Letters in Appl. Microbiol.*, 17, p.300-302.

Tzitzi, M., Vayneas, D.V., Lyberatos, G. (1994). Pretreatment of textile indsutrial wastewaters with ozone. *Wat.Sci.Technol.*, 29(9):151-160.

Vlyssides, A.G., Karlis, P.K., Rori, N., Zorpas, A.A. (2002). Electrochemical treatment in relation to pH of domestic wastewater using Ti/Pt electrodes. *Journal of Hazardous Matrials.*, B95:215-226.

Walling, C., Goosen, A. (1973). A mechanism of ferric ion catalyzed decomposition of hydrogen peroxide. Effect of organic substrates. *J. Am. Chem. Soc.*, 95(9):2987-2991.

Wang, G.S., Chen, H.W., Kang, S.F. (2001). Ctalyzed UV oxidation oforganic pollutants in biologically treated wastewater effluents. *The Science of the Total Environment*. 277:87-94.

Weber, E., Wolfe, N.L. (1987). Kinetic studies of reduction of aromatic azo compounds in anaerobic sediment/water systems. *Environ. Toxicol. Technol.* 6, 911-920.

Willmott, N.J. (1997). The use of bacteria-polymer composites for the removal of colour from reactive dye effluents. *PhD thesis, UK, Univeristy of Leeds.*

Wu, J., Eiteman, M.A., Law, S.E. (1998). Evaluation of membrane filtartion and ozonation processes for treatment reactive-dye wastewater. *J. Environ. Eng.*, 124:272.

Yang, Y., Wyatt, D.T., Bahrosky, M. (1998). Decolorization of dyes using UV/H2O2 photochmeical oxidation. *Textile Chemist Colorist.*, 30(4): 27-35.

Zappi, M.E. (1995). *Peroxone oxidation treatment of 2,4,6-Trinitrotoluene contaminated waters with and without sonolytic catalyzations.* Dissertation submitted to the faculty of Mississippi State University in partial fullfillment of the requirements for the degree of Doctor of Philosophy in chemical engineering in the department of chemical engineering..

Zhang, F.M., Knapp, J.S., Tapley, K.N. (1999). Decolourisation of cotton bleaching effluent with wood rotting fungus. *Wat.Res.*, 33(4), 919-928.

Zouari, N. (1998). Decolourization of olive oil mill effluent by physical and chemical treatment prior to anaerobic digestion. *J. Chem. Technol. Biotechnol.*73:.297-303.

In: Water Pollution: New Research
Editor: A. R. Burk, pp. 119-142

ISBN: 1-59454-393-3

Chapter 6

ADSORPTION OF IMIDACLOPRID BY HEAT TREATED KEROLITES: EFFECTS OF IONIC STRENGTH AND TEMPERATURE

E. González-Pradas*[1], *M. Socías-Viciana, M.D. Ureña-Amate and M. Saifi
Department of Inorganic Chemistry, University of Almería,
La Cañada San Urbano s/n. 04120-Almería, Spain

ABSTRACT

The adsorption of [1-(6-chloro-*3*-pyridinylmethyl)-*N*-nitro-2-imidazolidin-mine] (imidacloprid) on heat treated kerolite samples at 110°C (K-110), 200°C (K-200), 400°C (K-400) and 600°C (K-600) from pure water solution at 25°C has been studied. In addition, the influence of the presence of 0.1 M KCl in the medium and the influence of temperature, by performing the experiments at 10°C and at 40°C, was also investigated for the sample K-600. The evolution of surface properties of kerolite samples such as specific surface area and porosity after heat treatment were analysed. The clays were characterised by using: FTIR spectroscopy, XRD diffraction, TG and DTG analysis, surface analysis and Hg porosimetry. The experimental adsorption data points were fitted to the Freundlich equation in order to calculate the adsorption capacities (K_f) of the samples; K_f values range from 242 mg kg^{-1} for the K-110 sample up to 1005 mg kg^{-1} for the K-600 sample, that this, the stronger heat treatment, the more effective the adsorption of imidacloprid from pure water. The presence of 0.1 M KCl in the medium tends to increase slightly the adsorption of imidacloprid in the range of experimental conditions used. The variation of temperature did not show a clear influence on adsorption of imidacloprid from both, pure water and 0.1 M KCl solutions, since the adsorbed amount of imidacloprid seems to be dependent from the imidacloprid initial solution concentration (C_0). Nevertheless, the adsorbed amounts of imidacloprid at 25°C and 40°C are, in general, higher than those corresponding to 10°C.

[1] Corresponding author. Tel.: +34950015312, fax: +34950015008, e-mail: egonzale@ual.es

The results obtained from this work could be of interest by showing the better experimental conditions for the potential use of heat-treated kerolites -especially that activated at 600°C- for adsorption of imidacloprid. It seems, in general, that by conducting the adsorption process at 25°C in presence of 0.1 M KCl solution, the optimum adsorption of imidacloprid could be achieved, independently of the initial solution concentration of imidacloprid.

Key words: removal, insecticide, pollution, clay

INTRODUCTION

Imidacloprid is a systemic, chloro-nicotinyl insecticide effective for controlling sucking insects and used in soil and foliar treatment in different crops. It was introduced in the market with the commercial names Admire, Gaucho or Confidor. By virtue of its good contact properties and powerful systemic action after uptake through the root system, imidacloprid can be applied to soil and used as a seed dressing. This chloronicotinyl insecticide acts on cholinergic receptors of an identified insect motor neurone [1,2]. Owing to its insecticidal effectiveness at low concentrations and its safety for the environment, imidacloprid has drawn attention as a promising new comer [3]. In spite of that, accumulative effects can take place by successive treatments in soils, surpassing the safety margin.

Imidacloprid, according to the US Environmental Protection Agency (EPA), has the potential to leach to ground water. In addition, high solubility and mobility are concerns for transport to surface water through dissolution in runnoff [4]. Details about these concerns include the following: the half-life of imidacloprid in soil is 48-190 days, depending on the amount of ground cover, breaking down faster with plant ground cover than in fallow soils.

In addition, the ability of imidacloprid to move in soil has been demonstrated by a variety of studies, so it is considered by the United States Environmental Protection Agency (U.S.E.P.A.) to be a potential water contaminant [4-6]. Therefore, it would be desirable to find a single method to remove imidacloprid from surface and ground waters as well as to eliminate it from contaminated aqueous effluents.

As a result of several mineralogical studies in the last 10 years, it has been confirmed the existence of a new group of deposit formed by kerolite/stevensite interstratified material in the Tertiary basin of Madrid (Spain) [7]. Kerolite [$Mg_3Si_4O_{10}(OH)_2.nH_2O$] is a 2:1 magnesium silicate chemically and structurally similar to talc, although it is more disordered and has a higher degree of hydration [8-10]. The kerolite/stevensite material belonging to the Madrid basin is a kerolite-rich material (80% of kerolite) (named as kerolite in the text) and is of a special interest not only because it is used as a bleaching material but also because its potential application for the decontamination of organic molecules [7, 11-13].

Adsorption phenomena on solid substrates such as clays or activated carbon, is a widely used method for removing organic contaminants from waters [14, 15].

Previous studies have also demonstrated that adsorption can be affected by several parameters such as the temperature of the process and the presence of salts in the medium because of a possible competition of the organic molecules for the adsorption sites [16, 17].

Taking into account all the above indicated, it is useful to study the adsorption process of imidacloprid from pure water solution at 25°C on kerolite samples heat-treated at 110°C,

200°C, 400°C and 600°C, since previous research on the use of inorganic solids in removing insecticides has shown that the physico-chemical activity in the surface of some of these adsorbents is notably increased by means of heat-treatment [18, 19]. In addition, the influence of the presence of 0.1 M KCl in the medium and the influence of temperature, by performing the experiments at 10°C and at 40°C, has been also investigated for the sample K-600, since preliminary studies showed that this sample was the most effective in removing imidacloprid from pure water solution.

The objectives of the present research were: (i) to study the evolution of the surface and textural properties of the kerolite after heat treatment, (ii) to evaluate the effectiveness of the kerolite treatment in removing imidacloprid from pure water solution over a range of experimental conditions and (iii) to evaluate the effects of temperature and ionic strength on the adsorption of imidacloprid, from aqueous solution, by using kerolite activated at 600°C.

EXPERIMENTAL

Characterisation of Clay

The clay used as adsorbent for this series of experiments was a kerolite/stevensite interstratified material (80% of kerolite) from the Madrid Basin (Spain) (crushed to a particle size less than 150 μm), previously studied by other authors [7].

Heat activations of the natural kerolite were carried out by heating the sample at 110°C, 200°C, 400°C and 600°C for 24h. The four products so obtained are labelled in the text as K-110, K-200, K-400 and K-600, respectively. The materials obtained were analysed chemically by X-ray fluorescence spectroscopy with a Philips PW 1404 instrument. Specific surface area and micropore volume were determined from N_2 adsorption isotherms at 77.4 K, in a volumetric adsorption system, Gemini II-2375 (Micromeritics). The nitrogen used was 99.998% pure. The samples were degassed previously at 110°C for 24 hours.

Further information on the pore texture of the materials used, including contribution of the meso- and macropores to the total pore surface area and volume, was obtained using a Autopore III porosimeter (Micromeritics), with maximum pressures of 60000 Psi that covers spectra between 1.8 and 10000 nm. The system incorporates a data processor which facilitates analysis of the experimental extrusion and intrusion curves of mercury, pore size distribution, accumulative volume and equivalent surface area of pores. Thermogravimetry (TGA and DTG) curves were obtained with a Mettler TA 3000 thermogravimetric analysis unit and Mettler TG 50 thermobalance. The FTIR spectra of the samples were recorded using KBr pellets on an ATI Mattson spectrometer over a range of 4000-400cm^{-1}. The XRD patterns were obtained on a Philips PW-1700 diffractometer using graphite-monochromated Cu K_α radiation.

Sorption Experiments

Analytically pure imidacloprid was obtained from Bayer Hispania Industrial (Spain), and used as adsorbate in this study. Sorption isotherms were determined by batch equilibration of

0.25 g of the clay sample with 0.025 L of pure water solution (K-110, K-200, K-400, K-600) or 0.1 M KCl solutions (K-600) of imidacloprid of varied initial concentration (2.0 – 250 mg L^{-1}). Experiments were carried out in a thermostatic shaker bath at 25°C (K-110, K-200, K-400, K-600) or at 10°C, 25°C and 40°C (K-600). Preliminary experiments were conducted for various time intervals to determine when sorption equilibrium was reached. The sorption equilibrium time required for imidacloprid was 24 h.

After equilibration, the suspensions were centrifuged, the supernatant filtered and the pesticide concentration measured in the supernatant solution (C_e) by high performance liquid chromatography (HPLC) using a diode array detector and data station. Separation by isocratic elution was performed on a 150 mm x 3.9 mm id Waters Nova Pack bonded-phase column (C_{18} reversed-phase, particle size, 4 μm). The mobile phase was acetonitrile + water (65 + 35 by volume) and the pesticide detected at 270 nm. The amount adsorbed (X) was calculated from the difference in concentration between initial (C_i) and final (C_e) solution. Blanks containing no imidacloprid and two replicates of each adsorption point were used for each series of experiments. The pH of the blanks as well as those corresponding to the several solutions in contact with the kerolite used as adsorbent, were measured using a pH-meter Crison, model pH 2002. The pH values of the blanks containing no imidacloprid were 7.52 and 7.32 in pure water and 0.1 M KCl solution, respectively. No significant differences relating these values were observed for the measured pH of the several solutions of imidacloprid in contact with the kerolite sample.

Solubility of imidacloprid in pure water and in KCl 0.1 M solutions, at 10°C, 25°C and 40°C, was also determined by equilibrating in a thermostatic shaker bath, at each work temperature, saturated solutions of imidacloprid with analytically pure imidacloprid. Once the highest concentration had been reached, and maintained constant for 2 weeks, the equilibrium of the dissolution process was supposed completed, and the solubility determined.

In addition, FTIR spectra of the different samples saturated with imidacloprid were carried out as follows: an aqueous insecticide solution (0.025 L) with a concentration corresponding to the maximum used in the adsorption experiment was added under continuous stirring to 0.25g of each adsorbent in 0.1 L conical flasks. These flasks were shaken for the equilibration at 25°C. Following the equilibration period, samples were centrifuged and the supernatant concentration measured. Another 0.025 L portion of the same insecticide solution was added and the process was repeated until the supernatant concentration was constant. Then, remaining solids were air dried and a portion (0.001g) diluted with KBr in an approximate ratio to 1: 50 and pressed into pellets for analysis [20].

RESULTS AND DISCUSSION

Characterisation of Clay

The chemical composition of the major elements in the natural kerolite and the heat-treated samples is indicated in Table 1. Elemental analysis of the samples shows that in all cases, the samples have a high MgO content relative to Al_2O_3, Fe_2O_3 and K_2O. The results obtained are in the range reported by other authors, and confirm the trioctahedral and magnesian character of the mineral [21].

Table 1.- Chemical analysis of the kerolite samples. Percentage of major oxides

Sample	SiO_2 (%)	Al_2O_3 (%)	Fe_2O_3 (%)	CaO (%)	TiO_2 (%)	MnO (%)	K_2O (%)	MgO (%)	Na_2O (%)	P_2O_5 (%)	L.O.I.[a] (%)
K-110	54.69	2.41	1.029	0.885	0.172	0.019	0.690	25.30	0.067	0.015	14.54
K-200	54.87	3.20	0.863	0.772	0.144	0.022	0.544	26.55	0.081	0.045	13.01
K-400	57.84	2.66	0.949	0.810	0.152	0.022	0.605	27.49	0.067	0.026	9.51
K-600	58.85	2.40	0.976	0.859	0.158	0.024	0.666	27.89	0.081	0.025	7.98

[a] Lost on ignition

Figures 1a and 1b show the experimental N_2 adsorption-desorption isotherms at 77.4 K on the different kerolite samples. As can be seen from the shape of the isotherms, they can be classified as type II of the Brunauer, Deming, Deming and Teller classification (B.D.D.T.), showing a hysteresis cycle of H-3 type which is usual in aggregates of lamellar particles with feeble cohesion and with open pore in fissure form [22].

Surface area (S) of the samples available to the N_2 molecules were calculated by applying the Brunauer, Emmett and Teller equation (B.E.T.) to the experimental data points obtained in the N_2 adsorption process. The values of the specific surface area obtained for the kerolite samples were: 255 $m^2\ g^{-1}$ (K-110), 254 $m^2\ g^{-1}$ (K-200), 248 $m^2\ g^{-1}$ (K-400) and 224 $m^2\ g^{-1}$ for the K-600 sample. In all cases, the correlation coefficients were higher than 0.999. The heat treatment produces, in general, a decrease in the surface area values which may be due to the closing of micropores as the temperature increases [19].

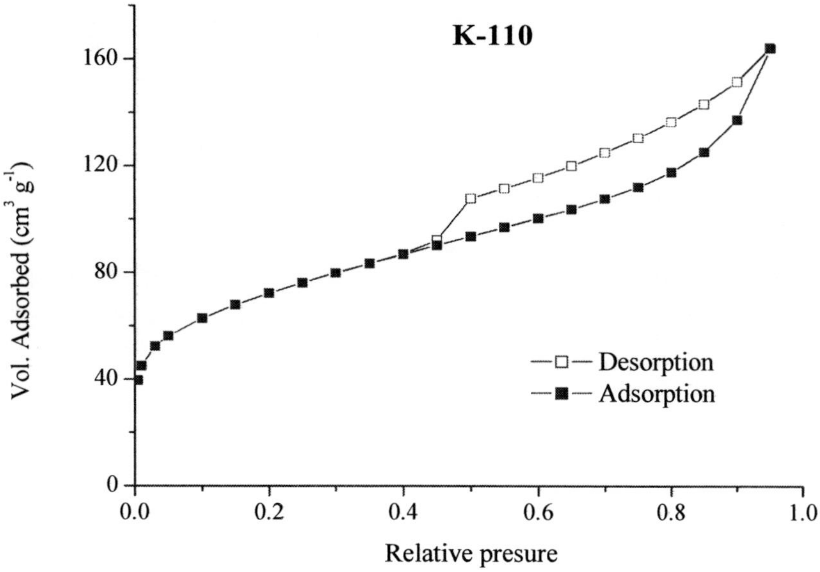

Figure continued on the next page

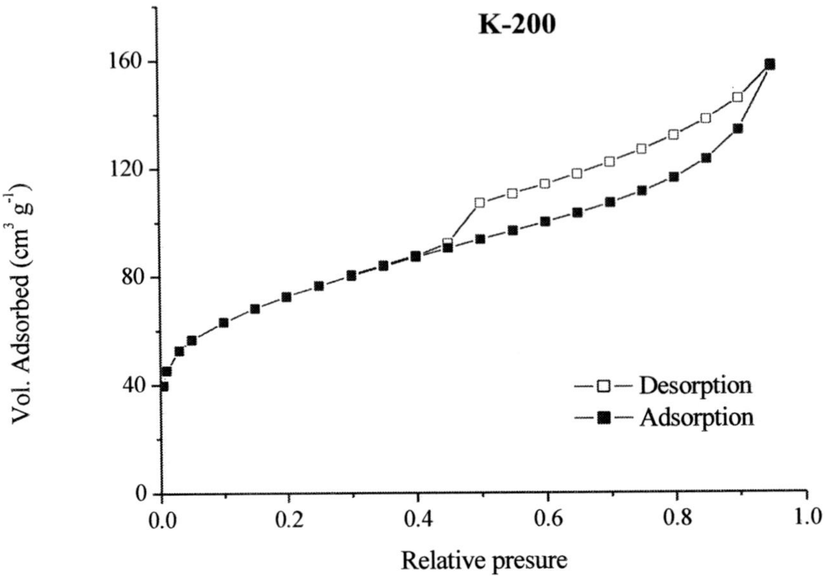

Figure 1a.- N_2 adsorption and desorption isotherms obtained for the heat treated kerolite samples

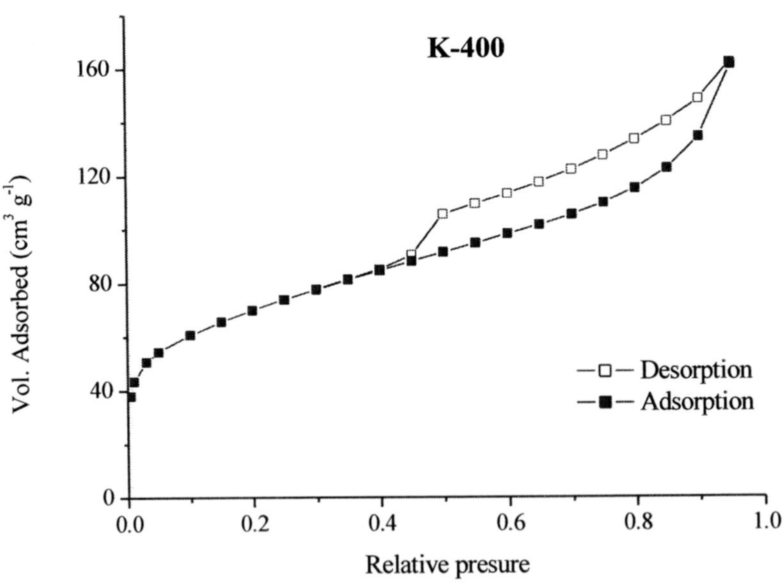

Figure continued on the next page

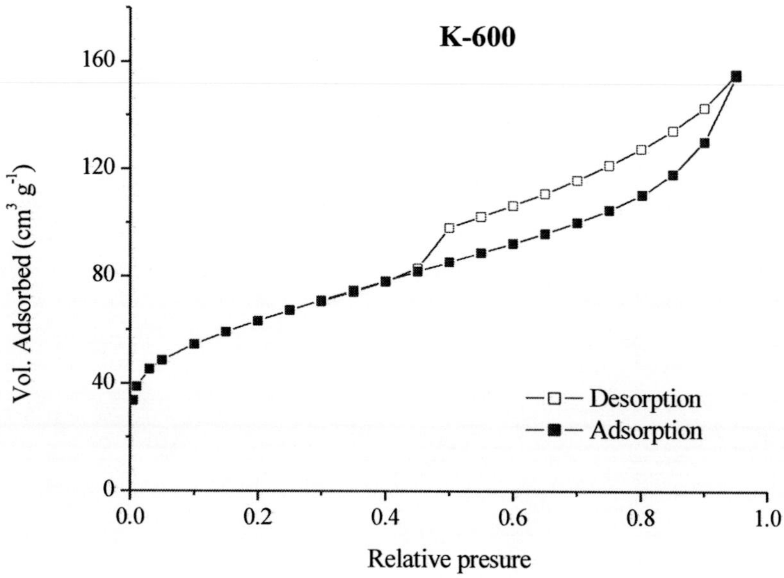

Figure 1b.- N_2 adsorption and desorption isotherms obtained for the heat treated kerolite samples

Table 2 compiles the values of the micropore volume [diameter (ϕ) $\leq$ 4 nm] obtained from analysis of the adsorption isotherms using the t-De Boer method [23] together with the meso (4 nm $\leq \phi \leq$ 50 nm) and macropore volumes (50 nm $\leq \phi \leq$ 8500 nm) obtained from the Hg porosimetry analysis. It can be seen that the heat treatment produces a decrease in the micropores volume of the samples, from 0.038-0.040 $cm^3 g^{-1}$ (K-200 and K-110) to 0.026 $cm^3 g^{-1}$ (K-600), while the corresponding meso and macropores volumes are not altered.

Table 2.- Micropores, mesopores and macropores volumes of the kerolite samples

Sample	$V^a_{micropores}$ ($cm^3 g^{-1}$)	$V^b_{mesopores}$ ($cm^3 g^{-1}$)	$V^c_{macropores}$ ($cm^3 g^{-1}$)
K-110	0.040	0.25	0.63
K-200	0.038	0.25	0.61
K-400	0.036	0.25	0.61
K-600	0.026	0.27	0.64

[a] Diameter (ϕ) $\leq$ 4 nm
[b] 4 nm $\leq \phi \leq$ 50 nm
[c] 50 nm $\leq \phi \leq$ 8500 nm

Figure 2 shows the pore-size distribution for meso and macropores (diameter greater than 4 nm) for the different samples. As can be seen from Figure 2, the major contribution corresponds to the macropore zone the maximum varying from 5000 nm (K-110) up to 10000 nm (K-600).

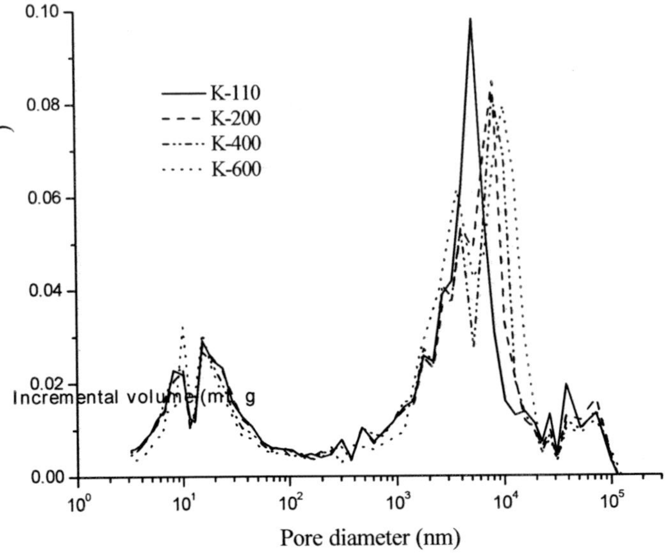

Figure 2.- Pore-size distribution curves for the heat treated kerolite samples

The thermogravimetric curves (TG) obtained for the kerolite samples are shown in Figures 3a and 3b. A continuous weight loss can be seen between 200°C and 800°C, the differential thermogravimetric curves (DTG) showing two endothermic effects. The first one is centered at about 130°C which corresponds to the loss of zeolitic water. The second one is centered about 800°C and is mainly associated to the loss of structural water which becomes evident from 400°C. These data are in agreement with those reported by other authors for similar samples [7].

Table 3 shows the total weight loss, the weight loss up to 400°C and the weight loss from 400°C to 850°C.

Table 3.- Percentage weight loss of the heat-treated kerolite samples

Sample	Weight loss (%) 35°C-400°C	Weight loss (%) 400°C-850°C	Total Weight loss (%)
K-110	5.78	6.33	12.11
K-200	4.73	5.18	9.91
K-400	2.23	5.02	7.25
K-600	2.48	4.16	6.64

As can be seen from Table 3, the total weight loss decreases, as expected, with the heat treatment given to the kerolite samples, from K-110 to K-600. The small weight loss observed at 130°C for the K-110 sample (5.78%) justifies according to Pozo et al. (1999), the major kerolitic character of the mineral used as adsorbent.

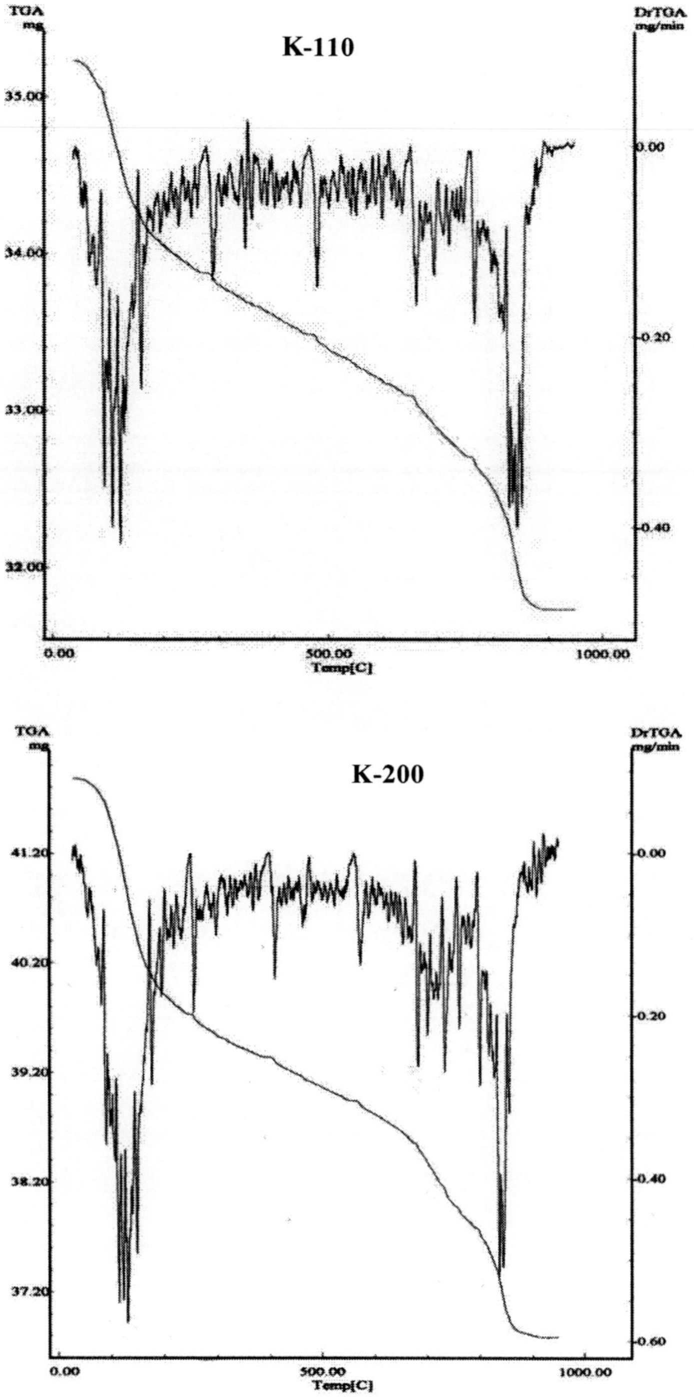

Figure 3a.- TG and DTG curves obtained for the heat treated kerolite samples

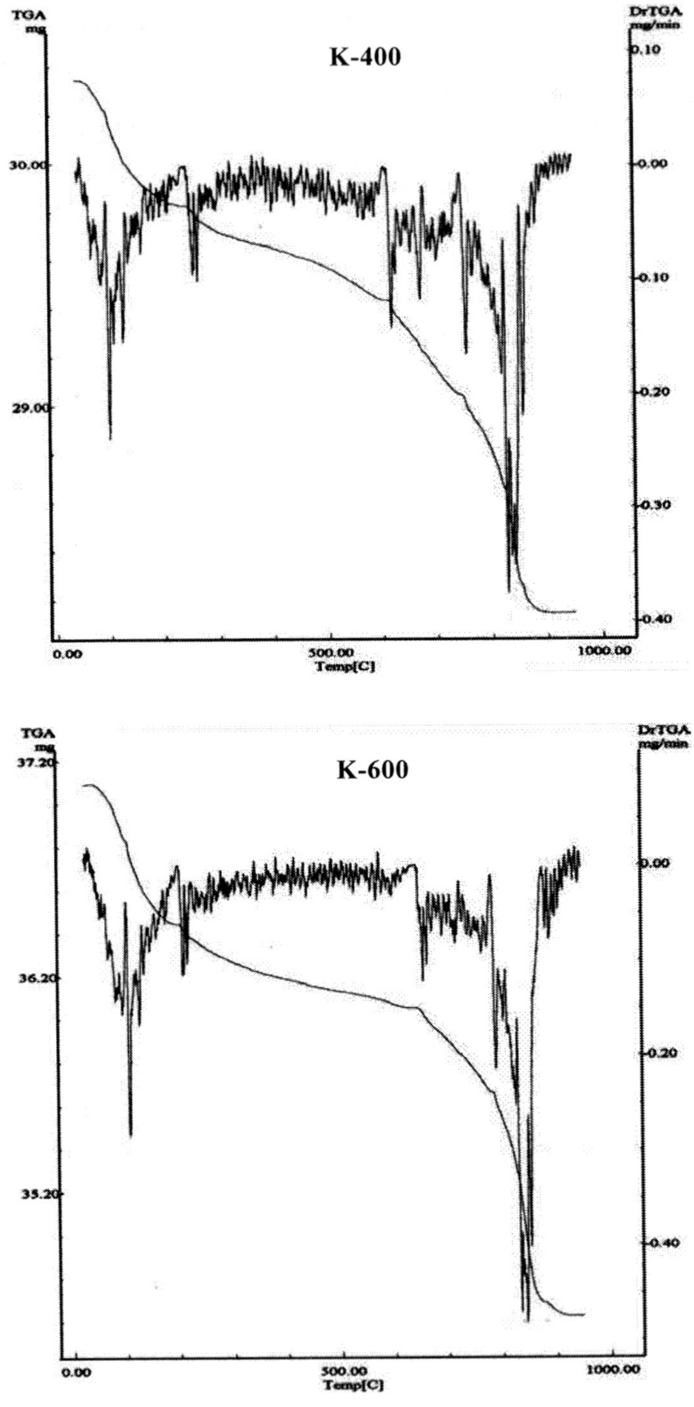

Figure 3b.- TG and DTG curves obtained for the heat treated kerolite samples

The FTIR spectra of the heat-treated kerolite samples are shown in Figure 4, where it can be observed the characteristic bands of this type of minerals, similar to those appearing in the literature [7,9,21,24].Thus it can be seen the following characteristic bands, centered at: a)

3440 cm^{-1}, corresponding to the O-H stretching associated with hydrogen bonds between interlayer water and surface oxygens; b) 1650 cm^{-1} corresponding to the vibrational mode of H-O-H; c) 1010 cm^{-1} assigned to the Si-O-Si stretching band and d) two bands centered at 660 cm^{-1} and 450 cm^{-1} corresponding to the vibration of the OH groups bounded to the octahedral Mg^{2+} and to the Mg-O bond, respectively. No significant differences were observed between the FTIR spectra obtained for the heat-treated samples.

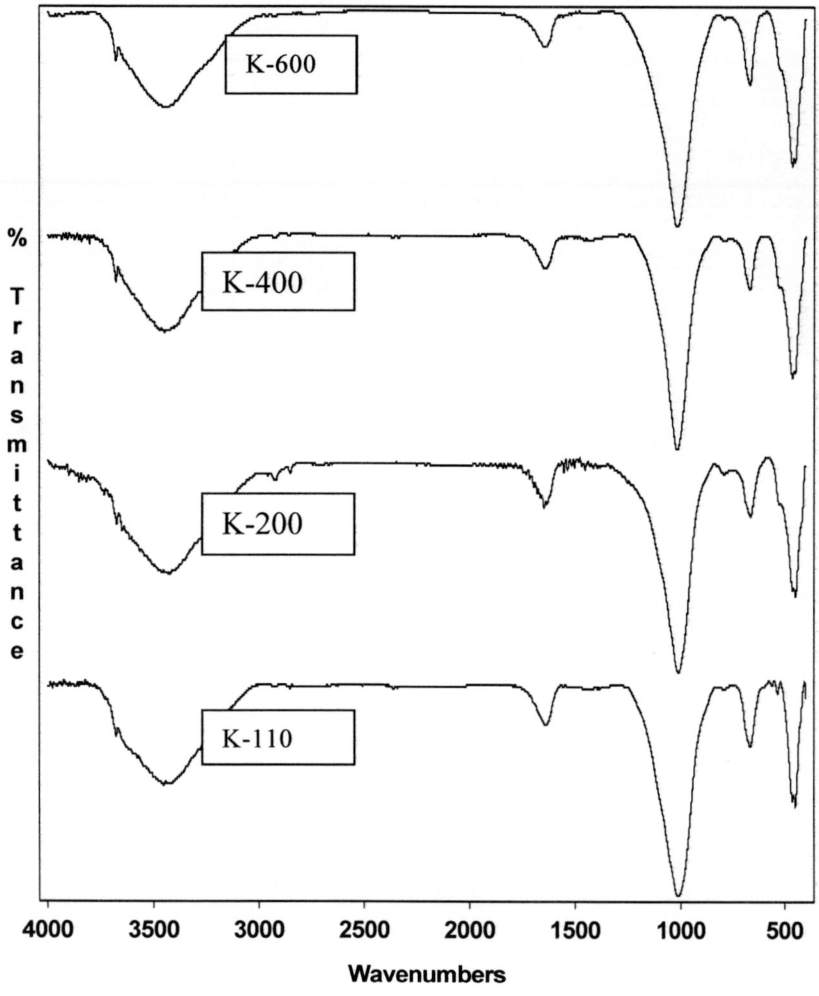

Figure 4.- FTIR spectra obtained for the heat treated kerolite samples

Figure 5 shows the X-ray diffraction patterns corresponding to the kerolite samples studied. The spacing sequences obtained in this study are similar to those reported for other kerolite-stevensite interestratified minerals [25]. The studied samples show a broad 001 reflection between 9.74 Å (K-600) and 10.08 Å (K-110) which is characteristic of those terms richer in kerolite among the kerolite-stevensite interestratified materials [7, 24]. The small decrease observed from 10.08 Å (K-110) to 9.74 Å (K-600) could be related to a small collapse of the structure as a consequence of the heat treatment [9]. On the other hand, it is also obtained the following spacing sequences: 4.53-4.57 Å (hk), 3.22-3.24 Å (003), 2.60-

2.62 Å (hk), 2.23-2.28 Å (hk), 1.73 Å (hk) and 1.53 Å (hk). It is interesting to note the strong and sharp reflection at 1.53 Å corresponding to the spacing d(060) which indicates the trioctahedral character of the mineral. The same reflection is observed for the heat-treated samples which shows that the trioctahedral character of the mineral is not affected by the heat treatment [7].

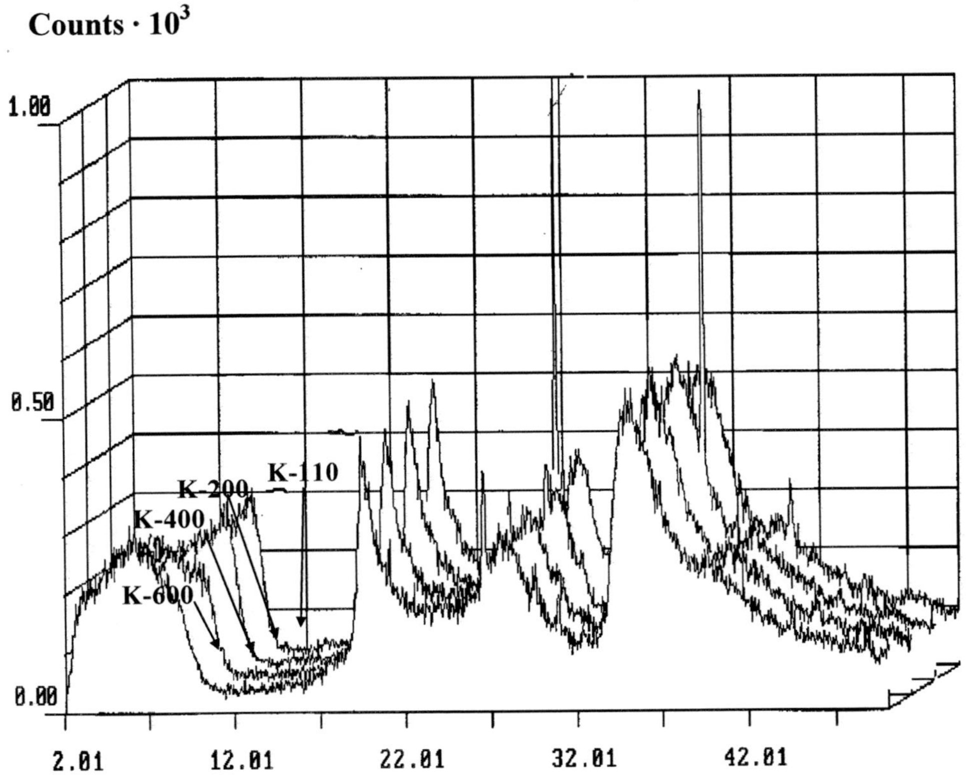

Figure 5.- X-ray diffraction patterns obtained for the heat treated kerolite samples

Sorption of Imidacloprid on the Heat-Treated Kerolite Samples

Figure 6 shows the adsorption isotherms of imidacloprid from pure water solution at 25°C on the kerolite samples (K-110, K-200, K-400, K-600). According to the slope of the initial portion of the straight lines obtained, these isotherms may be classified as C-type of the Giles classification [26]. This suggests that adsorption could take place as a partition of imidacloprid between the solution phase and the surface of the adsorbent. An increase of the slope of each straight line can be noted as the kerolite heat treatment increases, which is much more evident if the K-110 and K-200 samples are compared with the other heat-treated kerolites. This indicates an increased affinity of kerolite active sites for imidacloprid as the heat treatment increases. It is noteworthy that for a given equilibrium concentration, (C_e), the amount of imidacloprid adsorbed (X) is higher as the heat-treatment increases from 110°C to 600°C.

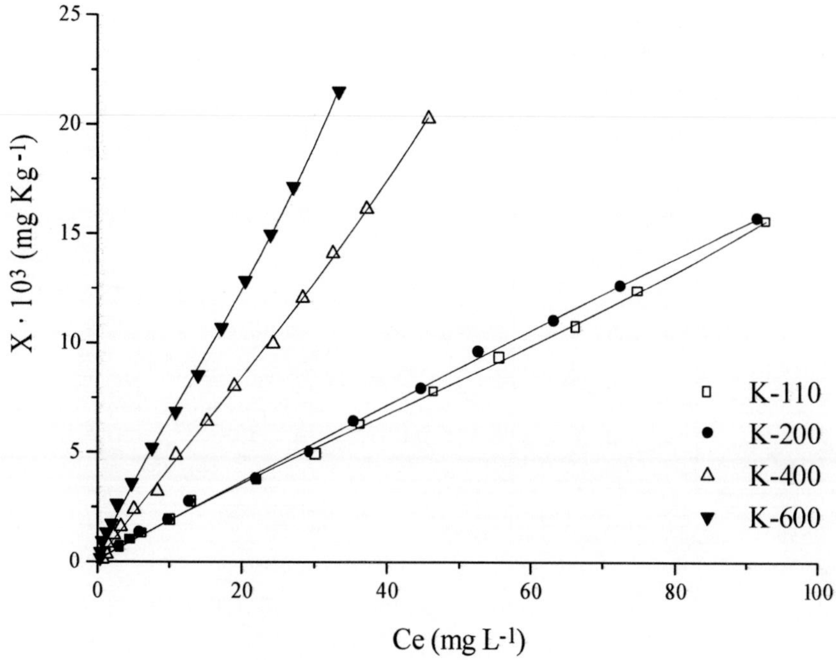

Figure 6.- Adsorption isotherms of imidacloprid on the heat treated kerolite samples

To discover the adsorption capacities of the kerolite samples, the experimental data points were fitted to the Freundlich equation [27]:

$$\text{Lg } X = \text{Lg } K_f + n\ \text{Lg} C_e \quad [1]$$

where

X = imidacloprid adsorbed per kilogram of kerolite, mg kg^{-1}
C_e = equilibrium solution concentration, mg L^{-1}
K_f= value of X at concentration equal to 1 mg L^{-1}
n = slope of the isotherms

Both, K_f and n values are constants being indicative of the extent of adsorption and the degree of non-linearity between solution concentration and adsorption, respectively [28]. Table 4 shows the K_f and n values calculated by the least square method applied to the straight lines obtained from equation [1]. The correlation coefficients were in all cases greater than 0.99 (all correlation were significant at the 0.001 probability level). As can be seen from Table 4, K_f values range from 242 mg kg^{-1} for the K-110 sample up to 1005 mg kg^{-1} for the K-600 sample; so, the heat treatment given to the kerolite greatly increases its adsorption capacity for the insecticide imidacloprid. The average value of K_f is about 300 times higher than that obtained for adsorption of imidacloprid on natural bentonite, 25 times higher than that obtained for the adsorption of this same pesticide on acid-treated bentonite (H_2SO_4 2.5 M) and 4 times higher than that obtained for the adsorption on humic acid [29].

Table 4.- Parameters of the Freundlich equation, X_{20} and removal efficiencies for adsorption of imidacloprid on the kerolite samples

Sample	K_f (mg kg^{-1})	n	X_{20} (mg kg^{-1})	R= $\left(\frac{C_0 - C_e}{C_0}\right) x100$ (% pesticide removed)
K-110	242 ± 0.23[a]	0.92 ± 0.003	3190	62.8
K-200	253 ± 0.20	0.90 ± 0.005	3534	63.1
K-400	574 ± 0.17	0.87 ± 0.005	8534	81.6
K-600	1005 ± 0.15	0.80 ± 0.004	12830	87.2

[a] These values represent the 95% confidence limits

On the other hand, in order to contrast quantities of imidacloprid adsorbed at a higher equilibrium concentration, X values at C_e = 20 mg L^{-1} (X_{20}) have been calculated from equation [1]. These values are also given in Table 4. The lowest and highest values of X_{20} also correspond to K-110 and K-600 samples, respectively. The stronger the heat treatment, the higher the adsorption capacity. The average values of the amounts adsorbed of imidacloprid at C_e= 20 mg L^{-1} are about 13 times higher than those calculated at C_e= 1 mg L^{-1}.

The removal efficiency, R, is defined as:

$$R = \left(\frac{C_O - C_e}{C_O}\right) \times 100 \qquad [2]$$

where C_O is the initial concentration (59.6 mg L^{-1}) of a pure water solution of imidacloprid (0.025 L) placed in a stoppered conical flask and shaken for 24 h with an amount (0.25 g) of the samples, and C_e is the equilibrium solution concentration; R is expressed in terms of percentages also in Table 4. As can be seen from Table 4, the removal efficiency values (R) range from 62.8 % for K-110 up to 87.2 % for K-600.

The fact that K_f, X_{20} and R increase as the heat treatment given to the kerolite increases, is probably due, according to the TG/DTG data above discussed, to a change in the kerolite structure owing to the loss of structural water, so allowing the imidacloprid molecules to have more sites for adsorption on the clay surface [18,19]. The adsorption mechanism could possibly take place by cation bridging which involves the formation of an inner-sphere complex between the kerolite Mg^{++} cations and the imidacloprid –NH group [30,31]. This might be confirmed from the FTIR spectra obtained for the kerolite samples saturated with imidacloprid. In Figure 7, the spectra for imidacloprid, K-110 and imidacloprid adsorbed on K-110 are shown as an example. As can be seen from Figure 7, two new bands appears at 1550 cm^{-1}and 1400 cm^{-1} in the spectrum of kerolite saturated with imidacloprid, which correspond to the stretching mode of the C=C and N=O bonds, respectively. The absence in this same spectrum of the band assigned to the N-H group present in the imidacloprid molecule (3400 cm^{-1}) seems to endorse our hypotheses about the proposed adsorption mechanism.

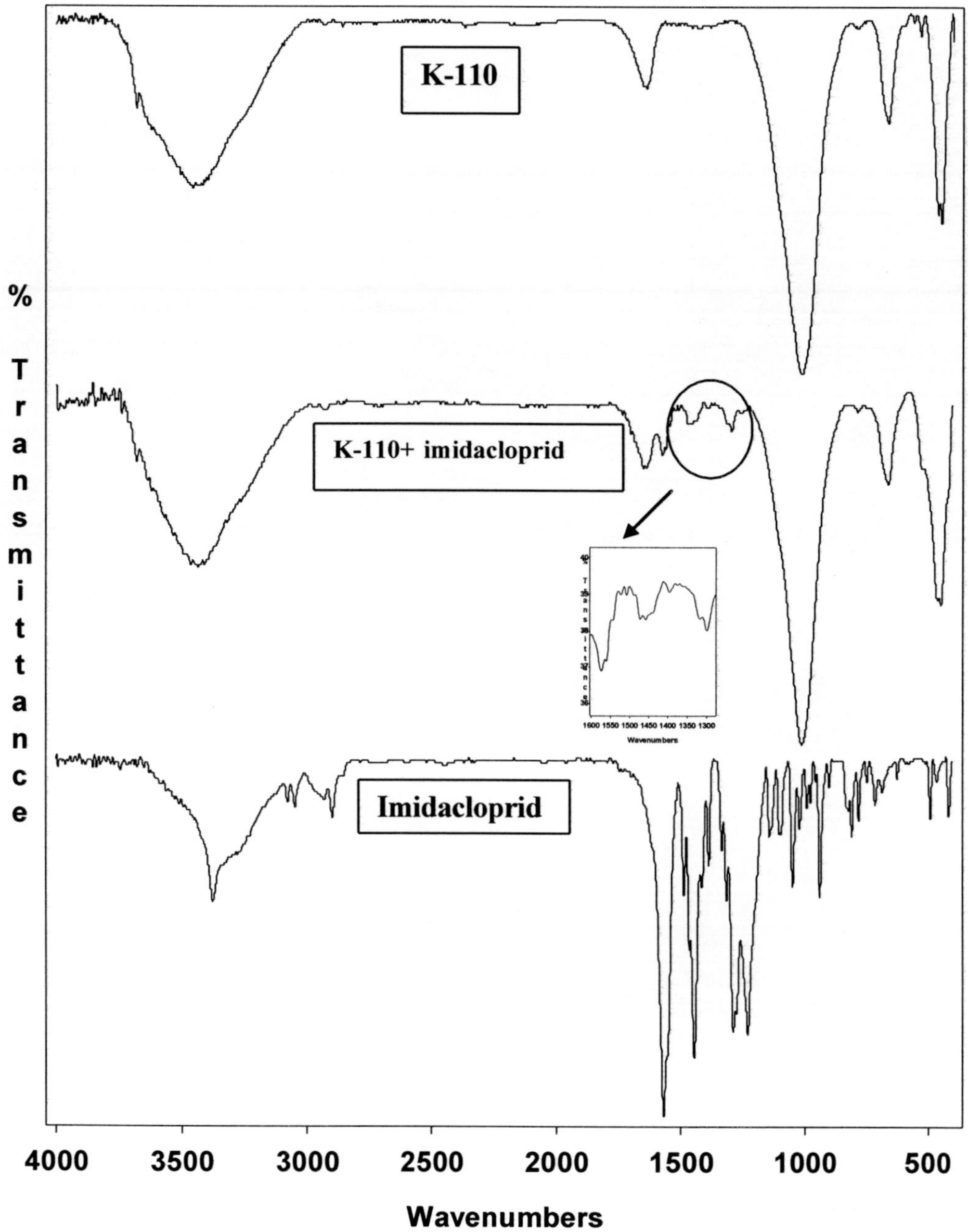

Figure 7.- FTIR spectra of imidacloprid, K-110 sample and imidacloprid adsorbed on K-110 sample

Effects of Ionic Strength and Temperature on Adsorption of Imidacloprid by the K-600 Sample

Figures 8a and 8b show the adsorption isotherms of imidacloprid on the K-600 kerolite sample (that showing the highest K_f value), at 10°C, 25°C and 40°C, in pure water and 0.1 M KCl solutions, respectively.

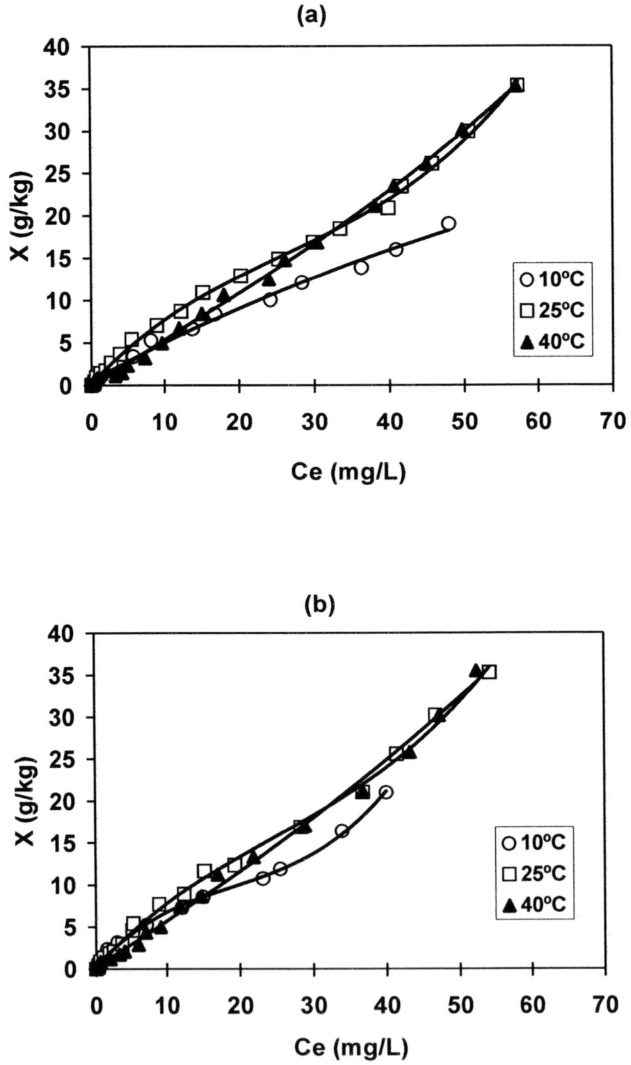

Figure 8.- Adsorption isotherms at 10°C, 25°C and 40°C for imidacloprid on the K-600 sample (a) in pure water medium (b) in 0.1 M KCl medium

According to the slope of the initial portion of the curves, these isotherms may be classified as S-type (40°C) or L-type (10°C and 25°C) of the Giles classification [26]. This suggests that kerolite has, respectively, a low or intermediate affinity for imidacloprid and that strong competition (40°C) or average competition (10°C and 25°C) from the solvent for adsorption sites occurs. It is also noteworthy in Figures 8a and 8b that the isotherms cross each other, so the influence of temperature does not appear clearly defined. The amount of imidacloprid adsorbed per kilogram of kerolite (X) at 10°C, 25°C and 40°C is highly dependent of the imidacloprid initial solution concentration (C_0). Thus, for C_e values from 0 to 15 mg L^{-1} the amount of imidacloprid adsorbed is maximum at 25°C and minimum at 40°C, whereas for C_e values between 15 mg L^{-1} and 35 mg L^{-1}, X is also maximum at 25°C but minimum at 10°C. Even more, for C_e values higher than 35 mg L^{-1}, X tends to be equal at 25°C and 40°C.

On the other hand, to analyse the effect of the ionic strength, Figure 9 jointly shows the adsorption isotherms of imidacloprid on the K-600 kerolite sample from both, pure water and 0.1 M KCl solution, at 10°C, 25°C and 40°C. As can be seen from Figure 9, the shape of the isotherms from 0.1 M KCl solution is the same than that for pure water solution. Nevertheless it is important to note that for a given equilibrium concentration, (C_e), the amount of imidacloprid adsorbed (X) is slightly higher when 0.1 M KCl is used as background solution in the adsorption experiments; the lower the temperature, the more important the influence of the ionic strength.

To evaluate the adsorption capacity of the 600°C heat-treated kerolite, at 10°C, 25°C and 40°C, in pure water and in 0.1 M KCl solutions, experimental data points were fitted to the Freundlich equation [27].

Table 5 shows the K_f and n values calculated by the least-squares method applied to the straight lines obtained from equation [1]. The correlation coefficients were in all cases greater than 0.99 (all correlation were significant at the 0.001 probability level).

Table 5. Parameters of the Freundlich equation, $X_{22.5}$ and X_{45}, for adsorption of imidacloprid on the K-600 sample

PURE WATER MEDIUM					
Temperature	K_f (g kg^{-1})	n	$X_{22.5}$ (g kg^{-1})	X_{45} (g kg^{-1})	r
10°C	0.84 ± 0.02[a]	0.80 ± 0.02[a]	10.2	17.7	0.9952
25°C	1.13 ± 0.02	0.82 ± 0.01	14.5	25.5	0.9968
40°C	0.32 ± 0.03	1.17 ± 0.02	12.2	27.4	0.9967
SALINE MEDIUM (0.1 M KCl)					
10°C	1.28 ± 0.03	0.72 ± 0.03	11.9	19.5	0.9967
25°C	1.17 ± 0.02	0.83 ± 0.02	15.3	27.2	0.9967
40°C	0.53 ± 0.02	1.04 ± 0.02	13.6	28.0	0.9973

[a] These values represent the 95% confidence limits

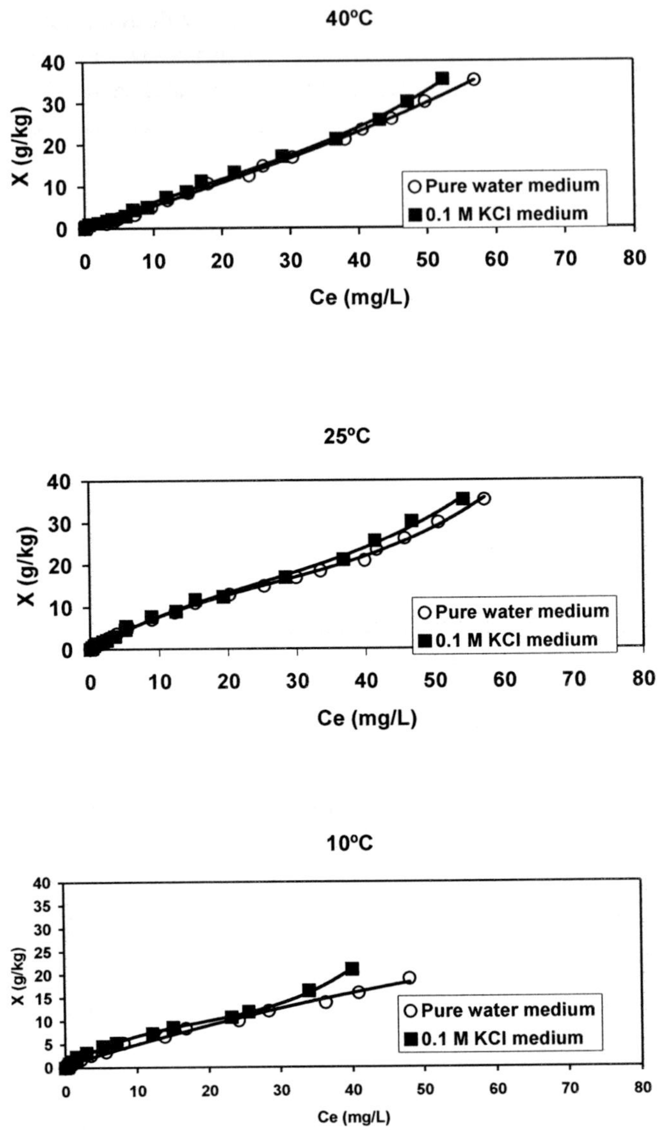

Figure 9.- Compared adsorption isotherms at 10°C, 25°C and 40°C (pure water and 0.1 M KCl solution) for imidacloprid on the K-600 sample

As can be seen from Table 5, K_f values range from 0.32 g kg^{-1} (40°C; pure water solution) up to 1.28 g kg^{-1} (10°C; 0.1 M KCl solution). The most prominent feature of these results is the trend observed for the K_f values as a function of the different temperatures studied. Increasing temperature from 10°C to 40°C results in a general decrease in the amount of imidacloprid adsorbed, this fact being clearer when the adsorption process is carried out in saline medium by using 0.1 M KCl as background solution. Nevertheless, as above indicated, the amount of imidacloprid adsorbed is clearly dependent of the imidacloprid initial solution concentration. So, in order to contrast quantities of imidacloprid adsorbed at equilibrium solution concentrations higher than 1 mg L^{-1} (value corresponding to K_f), we have considered it useful to calculate, from equation [1], the X values at C_e= 22.5 mg L^{-1} ($X_{22.5}$) and at C_e= 45 mg L^{-1} (X_{45}). The results obtained are also given, together with those corresponding to K_f, in Table 5. As can be seen from Table 5, $X_{22.5}$ and X_{45} values show a different trend than those corresponding to K_f, that is, increasing temperature from 10°C to 40°C results in a general increase in the amount of imidacloprid adsorbed, although the $X_{22.5}$ and X_{45} values tend to be equal at 25°C and at 40°C. This fact is in agreement with those above indicated when the experimental isotherms [X = f (C_e)] were analysed. To explain this behaviour it is important to take into account the dual influence of temperature on both, imidacloprid solubility and solvation grade of imidacloprid. Imidacloprid solubility increases as temperature increases. This fact may have a negative effect on the adsorption extent of imidacloprid, which seems to predominate at lower equilibrium solution concentrations. On the other hand, as temperature increases, the solvation grade of imidacloprid decreases. This originates a better diffusion of the imidacloprid molecules through the kerolite surface, so the amount of imidacloprid adsorbed increases as temperature increases. This fact seems to predominate at the higher equilibrium solution concentrations, although at 25°C and 40°C these two different effects seem to be balanced.

The effect of the electrolyte on the amount of imidacloprid adsorbed is also presented in Table 5. The adsorption of imidacloprid by the heat-treated kerolite sample from the 0.1 M KCl solution was higher than that obtained in no saline medium; the lower the equilibrium solution concentration and temperature, the more important, in general, the effect of the 0.1 M KCl background solution.

In order to clarify the influence of temperature on the adsorption process, and given that the solubility of imidacloprid varies greatly with temperature (from 212 mg L^{-1} at 10°C to 1015 mg L^{-1} at 40°C, the presence of ionic strength not affecting significantly the solubility), the adsorption data - as the amount of imidacloprid adsorbed per kilogram of kerolite (X), - were plotted versus the solubility-corrected equilibrium solution concentration (C_e/S). The adsorption isotherms so obtained appear in Figures 10a and 10b respectively, for pure water and 0.1 M KCl solutions.

As can be seen from Figures 10a and 10b, the influence of temperature seems to be well defined. Thus, for a given (C_e/S) value, the amount of imidacloprid adsorbed increases, in general, as temperature increases, that is, $X_{10°C} < X_{25°C} < X_{40°C}$. This fact confirms, according to the above mentioned, the two opposite effects of temperature on the imidacloprid adsorption. By suppressing the effect of solubility -by referring the equilibrium solution concentration to this factor (C_e/S)-, the only one temperature effect is the decreasing solvation grade of imidacloprid, so increasing the adsorption of imidacloprid as temperature increases.

We have tried to quantify the above by fitting the isotherms plotted in Figures 10a and 10b to a linear like-Freundlich equation:

$$\lg I = \lg K_s + n_s \lg C_e/S \quad [3]$$

Where

I = imidacloprid adsorbed, g kg^{-1}

C_e = equilibrium solution concentration, mg L^{-1}

S = solubility of imidacloprid at each temperature studied, mg L^{-1}

K_s = value of I at C_e equal to solubility

n_s = slope

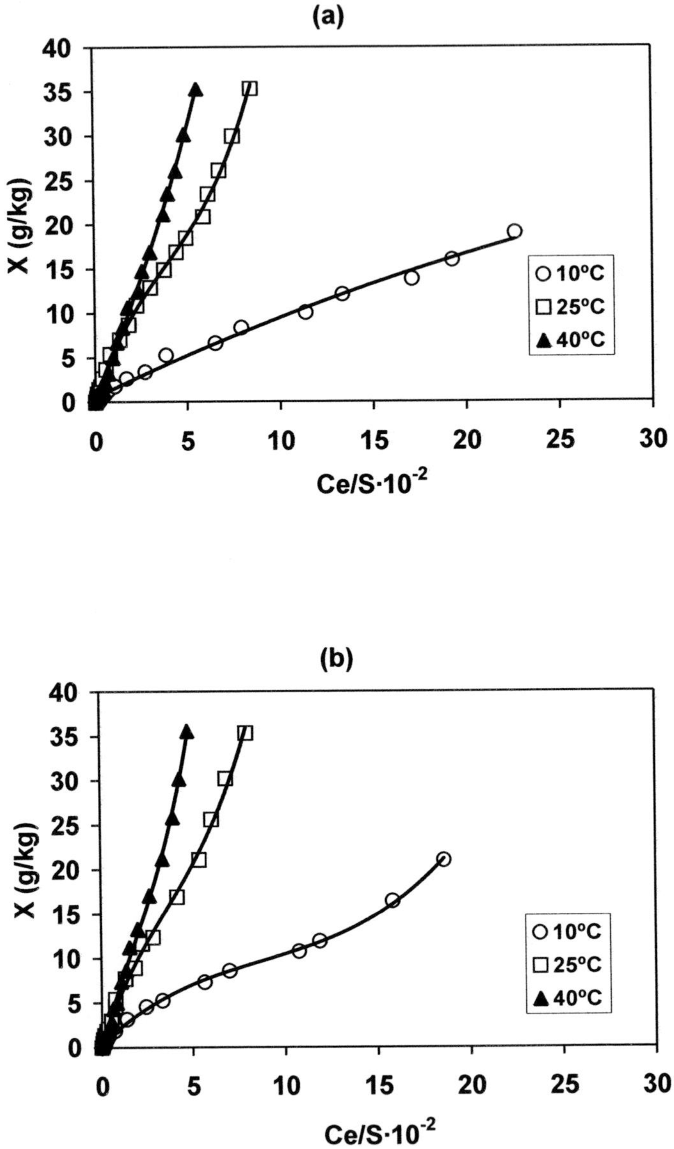

Figure 10.- Adsorption isotherms at 10°C, 25°C and 40°C for imidacloprid on the K-600 sample plotted in function of the solubility-corrected equilibrium solution concentration (a) in pure water medium (b) in 0.1 M KCl medium

Table 6 shows the K_s and n_s values calculated by the least-squares method applied to the straight lines obtained from equation [3]. The correlation coefficients were in all cases greater than 0.98 (all correlation were significant at the 0.001 probability level). From K_s and n_s values, the $I_{0.05}$ parameter has been calculated, which represents the amount of imidacloprid adsorbed at a $C_e/S = 0.05$. As can be seen from Table 6, $I_{0.05}$ values range from 5.6 g kg^{-1} (10°C; pure water solution) up to 27.8 g kg^{-1} (40°C; 0.1 M KCl solution). That is, increasing temperature from 10°C to 40°C results in a clear increase in the amount of imidacloprid adsorbed, from both, pure water and 0.1 M KCl solution. Likewise, the presence of the 0.1 M KCl tends to increase the amount of imidacloprid adsorbed.

Table 6. Parameters of a linear like-Freundlich equation and $I_{0.05}$ for adsorption of imidacloprid on the K-600 sample

Temperature	K_s (g kg^{-1})	n_s	$I_{0.05}$ (g kg^{-1})	r
PURE WATER MEDIUM				
10°C	61.6 ± 0.04[a]	0.80 ± 0.02[a]	5.6	0.9952
25°C	233.5 ± 0.03	0.82 ± 0.01	20.1	0.9969
40°C	348.9 ± 0.06	0.87 ± 0.03	25.8	0.9919
SALINE MEDIUM (0.1 M KCl)				
10°C	59.9 ± 0.06	0.72 ± 0.03	7.00	0.9867
25°C	256.1 ± 0.04	0.83 ± 0.02	21.5	0.9967
40°C	369.3 ± 0.08	0.86 ± 0.04	27.8	0.9858

[a] These values represent the 95% confidence limits

CONCLUSIONS

These experiments indicate that for all the heat treatments given to the kerolite, that carried out at 600°C is the most effective - at least in the experimental conditions used in this paper - in removing imidacloprid from pure water solution as shown by the higher adsorption and removal efficiency values.

The interest of the results obtained from this research work lies not only in the determination of the imidacloprid adsorption on heat treated kerolites but also in the potential application on pilot and industrial scale of this type of kerolite as a water decontaminant. The practical application of the adsorption process for the removal of imidacloprid from drinking water or waste water by using the heat treated kerolite here studied, could be accomplished in batch contact unit systems. In these systems, a predetermined amount of adsorbent is mixed with the water for a given contact period and subsequently is separated by sedimentation. The data obtained in this paper from the laboratory-scale equilibrium experiments – plotted according to the Freundlich equation – would provide information on the feasibility of using heat-treated kerolite to determine the theoretical capacity and usage rate of the kerolite to be applied for the removal of imidacloprid from environmental waters and drinking water resources.

The results obtained from this work could be of interest by showing the better experimental conditions for the potential use of the 600°C heat-activated kerolite for adsorption of imidacloprid. It seems, in general, that by conducting the adsorption process at 25°C in presence of 0.1 M KCl solution, the optimum adsorption of imidacloprid could be achieved, independently of the initial solution concentration of imidacloprid.

As this type of clay is relatively plentiful, this heat-activated kerolite sample might be reasonably used in order to remove imidacloprid from aqueous solutions, in the experimental conditions above proposed.

ACKNOWLEDGEMENTS

We thank SÜDCHEMIE for the kerolite samples

REFERENCES

[1] Bai D; Lummis SCR; Leicht W; Breer H; Sattelle DB. Actions of imidacloprid and a related nitromethylene on cholinergic receptors of an identified insect motor neurone. *Pestic. Sci.*, (1991) 33: 197-204.

[2] Liu MY; Latli B; Casida JE. Imidacloprid binding site in Musca nicotinic acetylcholine receptor: interactions with physostigmine and a variety of nicotinic agonists with chloropyridyl and chlorothiazolyl substituents. *Pestic. Biochem. Physiol.*, (1995) 52:170-181.

[3] Heuijbroeck W; Huijbregts AWH. Fungicides and insecticides applied to pelleted sugar-beet seeds.2. Control of pathogenic fungi in soil. *Crop Protection*, (1995) 14: 363-366.

[4] USEPA, Environmental Fate and Groundwater Branch. *EFGWB review of imidacloprid.* Washington, DC, Jun 11, 3 pp. (1993).

[5] Felsot AS; Cone W; Hu J and Ruppert JR. Distribution of imidacloprid in soil following subsurface drip chemigation. *Bull. Environ. Contam. Toxicol.* (1998) 60(3): 363-370.

[6] U.S. Geological Survey. *The quality of our nation's waters-nutrients and pesticides,* Circular 1225. Reston VA; USGS. 60 pp. (1999).

[7] Pozo M; Casas J; Martín de Vidales JL; Medina JA; Martín-Rubí JA. Características texturales y composicionales en depósitos de arcillas magnésicas de la Cuenca de Madrid. I) Kerolitas (sector de Esquivias y Pinto). *Boletín Geológico Minero* (1999) 110(1): 77-102.

[8] Maksimovic Z. β-Kerolite pimelite series from Goles Mountain, Yugoslavia. *Proc. Int. Clay Conf.*, Jerusalem, (1966) 1: 97-105.

[9] Brindley GW; Bish DL and Hsieng-Ming, W. The nature of kerolite, its relation to talc and stevensite. *Mineralogical Magazine* (1977) 41: 443-452.

[10] Zelazny, LW and White GN. The pyrophyllite-talc group, in *Minerals in Soil Environments, SSSA Book Series*, ed by Dixon J and Weed SB, n° 1, 2nd edn, Soil Science Society of America, Madison, WI, USA, (1989) pp. 527-550.

[11] González-Pradas E; Villafranca-Sánchez M; Socias-Viciana M; Cantos-Molina A; Ureña-Amate MD. Removal of chloridazón from water by kerolite/stevensite and bentonite: a comparative study. *J. Chem. Technol. Biotechnol.* (2000) 75: 1135-1140.

[12] González-Pradas E; Socias-Viciana M; Saifi M; Ureña-Amate MD; Flores-Céspedes F; Fernández-Pérez M; Villafranca-Sánchez M. Adsorption of atrazine from aqueous solution on heat-treated kerolites. *Chemosphere* (2003) 51: 85-93.

[13] Socias-Viciana, M.; González-Pradas, E.; Saifi, M.; Ureña-Amate, MD; Flores-Céspedes, F.; Fernández-Pérez, M. The potential removal of imidacloprid from water by heat-treated kerolites. *Pest Manag. Sci.* (2003) 59: 1162-1168.

[14] Svoboda, LP; Jandera P; Churácek J. Study of the sorption of selected pesticides on nonpolar sorbents. *Collect Czech. Chem. Commun.* (1991) 56: 317-333.

[15] González-Pradas E; Villafranca-Sánchez M; Gallego-Campo A; Ureña-Amate MD; Fernández-Pérez M. Removal of atrazine from aqueous solution by natural and activated bentonite. *J. Environ.Qual.* (1997) 26: 1288-1291.

[16] González-Pradas E; Villafranca-Sánchez M; Del Rey-Bueno F; Ureña-Amate MD; Fernández-Pérez M. Removal of atrazine and deisopropyatrazine from water by montmorillonite-phosphate crosslinked compounds. *J. Environ. Qual.* (2000) 29: 1316-1322.

[17] González-Pradas E; Villafranca-Sánchez M; Del Rey-Bueno F; Ureña-Amate MD, Fernández-Pérez M. Removal of paraquat and atrazine from water by montmorillonite-(Ce or Zr) phosphate crosslinked compounds. *Pest Manag. Sci.* (2000) 56: 565-570.

[18] González F; Pesquera C; Blanco C; Benito J; Mendioroz S and Pajares, JA. Structural and textural evolution under thermal treatment of natural and acid-activated Al-rich and Mg-rich palygorskites. *Appl. Clay Sci.* (1990) 5: 23-36.

[19] González-Pradas E; Villafranca-Sánchez E, Villafranca-Sánchez M; Del Rey-Bueno F; Valverde-García A and García-Rodríguez A. Evolution of surface properties in a bentonite as a function of acid and heat treatments. *J. Chem. Tech. Biotech.* (1991) 52: 211-218.

[20] Zussman J. *Physical methods in determinative mineralogy*. Academic Press, New York (1967).

[21] Martín de Vidales JL; Pozo M; Alía JM; García-Navarro F and Rull F. Kerolite-stevensite mixed layers from Madrid Basin, Central Spain. *Clay Miner.* (1991) 26: 329-342.

[22] Sing KSW; Everett DH; Haul RAW; Moscou L; Pierotti RA; Rouquerol J and Simienwska T. Reporting physisorcion data for Gas/Solid systems. *Pure and Appl. Chem.* (1985) 57: 603-619.

[23] Lippens BC; Linsen BG and De Boer JM. Pore systems in catalysis (I) adsorption of nitrogen: apparatus and calculation. *J. Catalysis* (1965) 3: 32-37.

[24] Pozo M. and Casas J. Origin of kerolite and associated Mg clays in palustrine-lacustre environments. The Esquivias deposit (Neogene Madrid Basin, Spain). *Clay Minerals* (1999) 34: 395-418.

[25] Martín de Vidales JL; Pozo M; Medina JM; Leguey S. Formación de sepiolita-paligorskita en litofacies lutítico-carbonáticas en el sector de Borox-Esquivias (Cuencas de Madrid). *Estudios Geol.* (1988) 44: 7-18.

[26] Giles CH; McEwan TH; Nakhwa SN and Smith D. Studies in adsorption. Part 2. A system of classification of solution adsorption isotherms. *J. Chem. Soc.* (1960) 4: 3973-3993.

[27] Adamson AW. Physical chemistry of surfaces. John Wiley and Sons, New York (1982).

[28] Grover R. Adsorption and desorption of urea herbicides on soils. *Can. J. Soil Sci.* (1975) 55: 127-136.

[29] González-Pradas E; Fernández-Pérez M; Villafranca-Sánchez M; Martínez-López F and Flores-Céspedes F. Use of bentonite and humic acid as modifying agents in alginate-based controlled-release formulations of imidacloprid. *Pesticide Science* (1999) 55: 546-552.

[30] Farmer VC and Mortland MM. An infrared study of the coordination of pyridine and water to exchangeable cations in montmorillonite and saponite. *J. Chem. Soc. A*, (1966) 344-351.

[31] Sposito G. *The surface chemistry of soils*. Oxford Univ. Press, New York (1984).

In: Water Pollution: New Research
Editor: A. R. Burk, pp. 143-159

ISBN: 1-59454-393-3

Chapter 7

APPLICATION OF AN "O_2/CO_2 HEADSPACE TEST WITH GC-TCD" FOR BIODEGRADABILITY TESTING OF SYNTHETIC ESTER LUBRICANTS IN WATER AND SOIL

Stefan Hahn[1] and Adolf Eisentraeger[2]
[1]Fraunhofer Institute of Toxicology and Experimental Medicine, Department Chemical Risk Assessment, Hannover, Germany
[2]University Hospital Aachen, Institute of Hygiene and Environmental Medicine, Aachen, Germany

ABSTRACT

A ready biodegradability test system is established which is based on the quantification of oxygen and carbon dioxide in the head space of batch vessels using gaschromatography with thermal conductivity detector. The test system is in accordance with a standard procedures ISO 10708 (BODIS-Test), ISO 11266 and ISO 14593 (ISO CO_2 Headspace Test) with slight modifications. It is applicable for the assessment of ready biodegradability in water and soil.

In this study the "O_2/CO_2 Headspace Test with GC-TCD" is used for the assessment of synthetic ester lubricants. The effects on biodegradability in water and soil of both additives and usage are examined and discussed. Ester based cutting fluids, gear fluids and hydraulic fluids with and without additives are used under defined conditions in machine tools and hydraulic and plain bearing test benches. The lubricants are characterised additionally with respect to kinematic viscosity, acidity and elemental composition. Furthermore, a formulated mineral oil is characterised before and after usage in the hydraulic test bench.

[1] Fraunhofer Institute of Toxicology and Experimental Medicine, Department Chemical Risk Assessment, Nikolai-Fuchs-Str. 1, 30625 Hannover, Germany, e-mail: hahn@item.fraunhofer.de
[2] University Hospital Aachen, Institute of Hygiene and Environmental Medicine, Pauwelsstr. 30, 52074 Aachen, Germany, e-mail: adolf.eisentraeger@post.rwth-aachen.de

Early results clearly show that the mineral oil is far less biodegradable than the ester oils and that their biodegradability is not affected by usage. The biodegradability of the ester oils mainly depends on the characteristics of the base fluids and is not affected by the additives. Antioxidants influence stability and biodegradability indirectly, since they prevent oxopolymerisation effects. Other effects of usage on biodegradation of lubricants are not detected. In this context, the antioxidants ensure ready biodegradability and have a positive effect on the environmental fate of synthetic ester lubricants. These statements can be affirmed in advanced studies with the results on biodegradability in soil. Furthermore, the fluids show a lower biodegradability in soil within 32 days than in water.

Key words: O_2/CO_2 Headspace Test with GC-TCD, ultimate aerobic biodegradation in water and soil, machine tools, hydraulics, plain bearing test bench

INTRODUCTION

13% (EC countries) and 32% (USA) of all lubricants used are emitted into the environment more or less unchanged in properties and appearance (Bartz, 1998). Most of these lubricants are used in loss lubrication frictional contacts and in circulation systems, in which the lubricants are disposed and not collected. Additionally, lubricants are emitted from leaks, and significant amounts remaining in filters and containers or disposals have to be taken into account (Bartz, 1998). Especially for hydraulic systems, which operate under high pressure. Therefore, large amounts of hydraulic fluids can be released into the environment from leaks or failures of hoses, seals or cylinders (Battersby, 2000).

For these reasons, the distribution, biodegradability and toxicity of lubricants are important factors with respect to a sustainable environmental development. Above all, the lubricants should not be persistent. On the other hand, the lubricants should be stable during usage under different operating conditions. Generally, lubricants change their chemical and physical fluid properties due to usage. Since the lubricant is a factor which considerably influences the economical as well as the ecological balance of a machine, the analysis of the aging behavior is extremely important. A characterization of the behavior of the properties of lubricants as a result of utilisation is thus the basis for the development of new lubricants.

The assessment of the environmental compatibility of chemical substances consists of exposure and effect assessment. Thereby, the effect assessment can be divided into aquatic und terrestrial effects and the exposure assessment comprises the releases into the environment and the distribution. The most important aspect with regard to the environmental fate of a substance is its biodegradability. It must be tested for e.g. the EU Dangerous Preparation Directive, eco-labels and the water hazard class classification (Willing, 1999, Willing, 2001). Furthermore, the biodegradability of a chemical is a part of the base data set of the new European chemical directive, which is called REACH concept (anonymous, 2003). Standardised methods are used for the testing ensuring comparable and reproducible results in different laboratories. These biodegradability test systems are described in norms such as OECD guidelines (OECD, 1992) and ISO methods (ISO/TR 15462, 1997). A review of these test systems is given by Pagga (1997). In the most cases, the so-called CEC-Test (CEC, 1995) is applied for lubricants. This is a test system, which was developed for two-stroke cycle outboard engine oils. However, biodegradability in soil is not routinely quantified although standardized protocols exist, e.g. ISO 11266 (1994). Furthermore, the bioavailability of the test substance can differ significantly under aquatic and terrestrial conditions in dependence on the physical-chemical properties of the substance.

First of all, the ingredients of lubricants have to be checked for their environmental compatibility. Generally, lubricants consist of base fluids and performance enhancing additives. Thereby, mineral oils, rapeseed oils, synthetic esters, native esters and other organic compounds are used as base fluids and the concentrations of additives such as antioxidants, extreme pressure additives and anti-wear additives are usually below 10% w/w. The biodegradability of the lubricants mainly depends on the biodegradability of the base fluid (Battersby, 2000). In fact, esters are readily biodegradable in contrast to mineral oils. For the eco-label "blue angel" it is tolerated, that the additives are only potentially degradable, if their concentration is below 7% w/w (RAL-UZ 79, 1996).

It is not generally recommended to test the biodegradability of used lubricants, although they alter significantly during usage. This may be caused by intake of metals, water or air as well as high pressure or temperature. The extent of the changes depends on the operating conditions. Examinations with the CEC-test have shown a decrease of biodegradability of synthetic ester lubricants during usage, although it still was superior to that of mineral oils (Schuelert and Bernhard, 1995). Remmele and Widmann (1998) have tested hydraulic fluids based on rapeseed oil in agricultural machinery with CEC-test and Zahn-Wellens-test. Only in a few cases biodegradability was not satisfying; in most cases it was extensive. However, the CEC-test and the Zahn-Wellens-Test, respectively, are not suitable for biodegradability testing of ester based lubricants.

Biodegradability of esters in soil has been demonstrated for triglycerides in laboratory experiments using three different soil samples (Hita et al., 1996). Field studies highlighted the potential of biodegradation of synthetic lubricants in soil (Haigh, 1995).

The effect of usage on biodegradability was examined basically by Eisentraeger et al (2002): the biodegradability of a synthetic hydraulic fluid without additives decreases after usage in a hydraulic test bench whereas the same hydraulic fluid with an additive package showed no decrease in biodegradability. In this work, a newly established test system according to ISO 10708 and ISO 14593 was used.

This "O_2/CO_2 Headspace Test with GC-TCD" is applied here for the characterization of the biodegradability of environmentally acceptable lubricants based on synthetic and oleochemical esters both in water and in soil. Effects of additives and usage are demonstrated. Furthermore, results of gear fluids are presented. These gear fluids were used in the same test benches as the hydraulic fluids.

MATERIALS AND METHODS

Chemicals and Lubricants

Analytical grade chemicals and the quartz sand were obtained from Merck KGaA (Darmstadt, Germany) and from Sigma-Aldrich Chemie GmbH (Steinheim; Germany). The tested lubricants for hydraulic, gear and cutting systems were commercially available. The base substances of the cutting fluid are di(2-ethylhexyl)adipate and isopropylpalmitate. The synthetic hydraulic fluid and the gear fluid contain trimethylolpropane-trioleate with native fatty acid composition as base substance. Besides, the hydraulic fluid contains di(2-ethylhexyl)adipate. The additive composition (antioxidants and extreme pressure/anti-wear substances) is unknown.

Test Benches

The degree of the changes due to usage of lubricants corresponds to the loads applied to the fluids. The investigation of aging thus has to take into account the various specific structures of the tribosystems. Since the investigations are carried out within the framework of the collaborative research centre SFB 442, tribosystems are regarded, which are used in a machine tool. So, fluids used in a hydraulic system, as well as in plain bearings (*Table 1a and 1b*), are analyzed with respect to the change of viscosity, the total acid number (TAN) and the elemental composition due to usage. The cutting fluid, which is used in the machine tool, is analyzed only with respect to the elemental composition.

The hydraulic fluid as well as the gear fluid are aged in a hydraulic circuit. In order to exactly simulate the loads applied to the fluids under real operating conditions the test stand is similar to industrially used stationary hydraulic systems. This means that a combination of different loads like temperature, oxygen as well as pressure are put on the fluid. Besides, the hydraulic circuit contains commonly used parts of typical industrially used hydraulic systems like a cooler, a tank and a filtration system. The flow circuit works as follows: The pump works against a pressure relief valve which leads to a pressure of 35 MPa. Behind the pressure relief valve, the temperature is about 120 °C. The fluid flow into the tank is 45 l/min. After passing a cooler, the fluid arrives again at the suction side of the pump. Blank samples of the fluids are collected immediately after filling the test bench and mixing ("base fluid blank EN1-0" and "with additives blank EN1-0").

Experiments with a plain bearing test bench are performed to evaluate the combination of fluid and materials. The hydraulic fluid as well as the gear fluid are used with bronze as slide bearing material (CuSn8, free of cadmium and arsenic). The fluid flows into the tank with 3 l/min. After passing a filter, the fluid arrives again at the plain bearing. The tank is heated to a temperature of 80 °C (AGL-1) or 90 °C (AGL-2) in the bearing. The pressure in the bearing is adjusted to 2.5 N/mm^2.

The cutting fluid is used in a machine tool. It belongs to the group of metalworking fluids and its general task is cooling as well as lubrication during cutting operation. It is responsible for the following aspects: accelerated heat dissipation with increased tool service life, lubrication between tool, workpiece and chips and improved chip removal. Fluids are analyzed after 5 hours and 30 hours of cutting operation with steel X5CrNi18-10 as work piece material and a tank volume of 200 l. A blank sample of the fluid with additives is collected immediately after filling the machine tool and mixing ("with additives blank ZDR-0").

Table 1a: Characterization of the tested fluids (Eisenträger et al. 2002)

	Parameter of usage	Kin. viscosity at 40 °C (mm^2/s)	Total acid number (mg KOH/g)
cutting fluids			
base fluid	unused	7.7	0.2
base fluid used ZDR	5 h, material: X5CrNi18-10, tank volume: 200 L	n. d.	n. d.
with additives	unused	7.7	3.0
with additives blank ZDR-0	0 h, material: X5CrNi18-10, tank volume: 200 L	n. d.	n. d.
with additives used ZDR-1	30 h, material: X5CrNi18-10, tank volume: 200 L	n. d.	n. d.
hydraulic fluids			
base fluid	unused	31.0	0.83
base fluid blank EN1-0	0.5 h, 90-120 °C, 2.5 kW/L, 35 MPa, open system	31.0	0.84
base fluid used EN1-1	308 h, 90-120 °C, 2.5 kW/L, 35 MPa, open system	459	31
with additives	unused	31.9	1.0
with additives blank EN1-0	2 h, 90-120 °C, 2.5 kW/L, 35 MPa, open system	32.3	1.2
with additives used EN1-1	308 h, 90-120 °C, 2.5 kW/L, 35 MPa, open system	91.5	11.4
with additives used EN1-2	312 h, 90-120 °C, 2.5 kW/L, 35 MPa, closed system	32.1	1.2
with additives used AGL-1	80°C, 1022 h, material: CuSn8	37	2.4
with additives used AGL-2	90°C, 832 h, material: CuSn8	59.3	4.5
mineral oil	unused	35.6	2.0
mineral oil used EN1-1	312 h, 90-120 °C, 2.5 kW/L, 35 MPa, open system	36.8	1.7

n. d. = not determined, ZDR = machine tool: cutting, EN1 = hydraulic test bench, AGL: plain bearing test bench

Table 1b: Characterization of the tested fluids

	Parameter of usage	Kin. viscosity at 40 °C (mm^2/s)	Total acid number (mg KOH/g)
gear fluids			
base fluid	unused	95.6	0.83
base fluid blank EN1-0	0.5 h, 90-120 °C, 2.5 kW/L, 35 MPa, open system	95.6	1
base fluid used EN1-1	312 h, 90-120 °C, 2.5 kW/L, 35 MPa, open system	434	11
with additives	unused	99	0.925
with additives used EN1-1	312 h, 90-120 °C, 2.5 kW/L, 35 MPa, open system	122	4.2
with additives used AGL-1	80°C, 1205 h, material: CuSn8	104	2.36
with additives used AGL-2	90°C, 1183 h, material: CuSn8	120	3.26

EN1 = hydraulic test bench, AGL: plain bearing test bench

Viscosity and Total Acid Number (TAN)

Kinematic viscosity is a measure of the relative flow of a fluid under influence of gravity. The laboratory determination of viscosity is carried out by using capillary tubes according to DIN 51562 (1978). The measurement is performed with a calibrated glass capillary viscometer, where the fluid passes through a fixed diameter orifice under the influence of gravity.

The TAN is measured according to DIN 51558 (1979) for the determination of acidic and basic constituents in petroleum products.

Combined Biodegradability Test System in Water

The batch assay called "O_2/CO_2 Headspace Test with GC-TCD" is performed according to standard procedures ISO 10708 (BODIS-Test) and ISO 14593 (ISO CO_2 Headspace Test) with slight modifications. It is based on the measurement of the ultimate aerobic mineralisation of test substances to carbon dioxide in water. Results are expressed as biological oxygen demand in relation to the theoretical oxygen demand (BOD/ThOD) and as carbon dioxide production in relation to the theoretical carbon dioxide production (TIC/ThIC). The theoretical values are calculated on the basis of values from elemental analysis (Elemental Analyzer Vario EL) in duplicate. The test substance is evaluated as "readily biodegradable", if degradation exceeds 60% within 28 days. The principle of the 10-d-window, which is prescribed in the OECD Guidelines, is neglected.

Test conditions are: 1 l infusion bottles closed with chlorobutyl stoppers (both: Zscheile and Klinger, Hamburg, Germany) and additional chlorobutyl septa (Macherey-Nagel, Dueren, Germany, N35B red); 800 ml mineral medium (ingredients as described in ISO 10708 (1997)); test substances (100 mg ThOD) are directly weighed in; stirring (600 1/min); room temperature (20 – 25 °C). Natural standard soil Lufa 2.2 (Landwirtschaftliche Untersuchungs- und Forschungsanstalt Speyer, Germany) is used as the inoculum instead of activated sludge. The soil characteristics are described in *Table 2*. 2 g soil/l correspond to approximately 10^7 cfu of bacteria and fungi per liter medium.

Total carbon dioxide production is determined after acidification of liquid aliquots with 0.1 M orthophosphoric acid (1 ml) at the end of incubation in duplicate. The measurements are performed in 7 mL test tubes closed with GL 14 hole-caps and silicone septa (Alltech GmbH, Unterhaching, Germany, blue septa ½"). The sample volume is 2 mL using a gastight syringe (SGE, Austin, Texas, USA, Type 2.5MDR-GT with valve SGE, Type VLL25/2.5). The recovery of CO_2 is between 95% and 105%.

Oxygen and carbon dioxide in headspace samples are measured gaschromatographically using a thermal conductivity detector (Perkin Elmer Autosystems) and a CTR 1 column (Alltech GmbH, Unterhaching, Germany). The sample volume of 100 µL is taken with a gastight syringe (Hamilton, Reno, Nevada, USA, Type 1710 SL). The analysis parameters are: oven temperature 40° C isothermal, detector temperature 140 °C and gas flow helium (purity grade 99.999 vol.%) 62.5 ml/min. The single measurements are taken at least weekly. Calibration is done with three test gases (Linde AG, Duesseldorf) of the following compositions (vol.%): test gas 1: CO_2 (0.02), O_2 (20.8), N_2 (79.18); test gas 2: CO_2 (0.8), O_2 (20.8), N_2 (78.4); test gas 3: CO_2 (5.0), O_2 (5.0), N_2 (5.0), He (85.0).

Table 2: Soil characterization of LUFA 2.2 (Eisenträger et al. 2002)

parameter	*unit*	*value*
kind of soil		loamy sand [a]
granulation size	(mm)	≤ 2 [a]
pH-value		5.8 ± 0.2 [a]
TOC	(%)	2.19 ± 0.08 [a]
water content	(%)	13.2
dry weight	(%)	86.8
WHC_{max}	(g / 100 g dw)	50 ± 4 [a]
cfu (fungi)	(cfu / g dw)	$1.1 \bullet 10^6$
cfu (bacteria)	(cfu / g dw)	$9.5 \bullet 10^6$
water extractable nitrite	(mg NO_2^- / kg dw)	< 1.0
water extractable nitrate	(mg NO_3^- / kg dw)	3.3
water extractable phosphate	(mg PO_4^{3-} / kg dw)	0.6
water extractable ammonium	(mg NH_4^+ / kg dw)	0.1

[a] specification LUFA, Speyer, Germany

Oxygen (appr. 100 hPa) is redosed with a 100 mL gastight syringe (SGE, Austin, Texas, USA, Type 100MR-VLLMA-GT), if the oxygen partial pressure in the incubation bottles falls below 100 hPa. This is performed once within 28 days for good biodegradable substances.

Test substances, reference (sodium benzoate) and blanks are tested in triplicate. The replicability of degradation is indicated as standard deviation taking into account the standard deviations of the blank-values and of the biological oxygen demand or carbon dioxide production and an element of uncertainty of the ThOD or ThIC. The assumed elements of uncertainty are: 0.1 g O_2/g for the ThOD, 0.01 g CO_2/g for the ThIC, and 0.01 g O_2/g for single substances, where the values are calculated from the molecular formula.

Combined Biodegradability Test System in Soil

The batch assay called "O_2/CO_2 Headspace Test with GC-TCD" can also be performed in soil according to the standard procedure ISO 11266. Thereby, the following modfications of the test procedure in water are necessary:

50 mg test substance are added to 2 g quartz sand in order to improve homogeneous distribution of this low amount of test substance in 50 g dry matter of standard soil Lufa 2.2 (which is used also as inoculum in the test system in water described above). The soil and the treated quartz sand are filled into 1 l infusion bottles (Zscheile and Klinger, Hamburg, Germany) and mixed. The water content of soil is adjusted to ca. 50% of the maximum water holding capacity. Nutrients like nitrogen or phosphorus do not have to be added, if Lufa 2.2 is used as soil sample. The 1 l infusion bottles are closed with chlorobutyl stoppers (Zscheile and Klinger, Hamburg, Germany) and additional chlorobutyl septa (Macherey-Nagel, Dueren, Germany, N35B red) like the test system in water. If the oxygen partial pressure in the incubation bottles falls during the incubation time below 100 hPa, the headspace is exchanged with test gase (Linde AG, Duesseldorf) (composition (vol.%): CO_2 (0.02), O_2 (20.8), N_2

(79.18)). All other test conditions except for stirring are similar to the conditions of the biodegradability test in water.

The assays are incubated for 120 days according to ISO 11266. However, in this work, the results after 32 days incubation are presented since they can be compared with the results of the biodegradability test in water. With the test system applied here, the complete mineralization of a complex substance is assessed by quantification of both oxygen and carbon dioxide turnovers. However, further endpoints are prescribed in ISO 11266. The primary degradation is not assessed, and C13-measurements are not performed since the used fluids are not available as C13-containing samples.

Elemental Analysis

The extent of biodegradation is expressed as a percentage of the theoretical O_2 uptake (ThOD) or CO_2 production (ThIC) for complete biodegradation of the test substance. These maximum values are normally calculated from the molecular formula. This is impracticable for most lubricants, because of their unknown chemical composition.

Elemental analysis is a usable method to calculate both ThOD and ThIC (Gerike, 1984 and Battersby, 2000), thus the theoretical values are calculated on the basis of values from elemental analysis (Elemental Analyzer Vario EL) in duplicates. Furthermore, the elemental composition demonstrates oxidative changes of the lubricants.

RESULTS

Viscosity and Total Acid Number

Viscosity is the measure of internal friction in a fluid. It is one of the most important properties for characterization of lubricants and their flow and transport behavior. The acidity of unused oils and fluids is normally derived from the type and concentration of the ester and specific additive materials, whereas the acidity of used oils is of interest to measure the degree of oxidation of the fluid. Both viscosity and TAN have been chosen to describe the degree of aging of the hydraulic fluids and gear fluids in routine.

In this work, the environmentally acceptable ester based oils were used in a hydraulic system as well as in a bearing test stand, accompanied by experiments with a mineral oil including additives in the hydraulic system. In contrast, the ester based oils were tested on the one hand as a base oil with no additives and on the other hand as a formulation incorporating additives. Viscosity and TAN of the used fluids are presented in *Table 1a and b*. First of all, the oxidation stability of mineral oil significantly exceeds the one of ester oils. Further, the investigations demonstrate that the tests age the fluids most of all, which are based on exposing the fluid to oxygen. Finally, results from plain bearing tests show the strong influence of temperature on the oxidation process (with additives used AGL-1 and AGL-2). Comparable results are obtained for the hydraulic and gear fluids. The viscosity after usage of 312 h is in the same order of magnitude for both fluids. Due to the higher initial viscosity of the gear fluids in comparison to the hydraulic fluids, the differences are marked in a lower degree between unused fluids and used fluids as well as the base fluids and the formulations.

Elemental Analysis

The results of elemental analysis for the tested lubricants are shown in *Table 3a and b*. The elemental compositions of all cutting fluids are equal except for the sample "with additives used ZDR-1" having a slightly decreased hydrogen content. No significant effects of usage or additives can be detected.

Table 3a: ThOD and ThIC of the fluids calculated via elemental analysis (CHN-Analyzer) (Eisenträger et al. 2002)

	C (%)	H (%)	O (%)[a]	ThOD (g O_2/g)	ThIC (g CO_2/g)
cutting fluids					
base fluid	72.9	12.4	14.6	2.80	2.67
base fluid used ZDR	72.8	12.1	15.1	2.76	2.67
with additives	72.8	12.3	14.9	2.78	2.67
with additives blank ZDR-0	72.5	12.0	15.5	2.74	2.66
with additives used ZDR-1	72.9	10.9	16.3	2.65	2.67
hydraulic fluids					
base fluid	75.9	11.2	12.9	2.79	2.78
base fluid blank EN1-0	76.1	11.8	12.2	2.85	2.79
base fluid used EN1-1	70.9	10.9	18.1	2.59	2.60
with additives	75.9	11.8	12.2	2.85	2.78
with additives blank EN1-0	75.2	12.3	12.5	2.87	2.75
with additives used EN1-1	72.6	11.9	15.5	2.73	2.66
with additives used EN1-2	76.4	12.0	11.6	2.88	2.80
with additives used AGL-1	75.8	11.6	12.6	2.83	2.78
with additives used AGL-2	74.8	11.5	13.7	2.78	2.74
mineral oil	85.0	13.6	1.4	3.35	3.11
mineral oil used EN1-1	84.9	13.4	1.6	3.32	3.11

[a] O (%) = 100 – (H (%) + C (%)); ZDR = machine tool: cutting, EN1 = hydraulic test bench, -0 = blank, -1 = open system, -2 = closed system, AGL = plain bearing test bench, -1 = 80 °C, -2 = 90 °C

The same is true for the hydraulic fluids with two exceptions: The hydraulic fluids "base fluid used EN1-1" and "with additives used EN1-1" show an increase of oxygen concentration of 12.2% up to 18.1% for the base fluid and of 12.5% up to 15.5% for the fluid with additives caused by the thermic-oxidative operating conditions. Taking into account also the increase of viscosity and the lower solubility in pentane, it can be concluded, that both fluids were partly oxopolymerised. The low oxygen concentrations of the mineral oils are probably caused by the additives. Due to the high oxygen content the ester oils have lower ThOD and ThIC values than the mineral oils.

The results of the gear fluids confirm the results, which are obtained with the hydraulic fluids. The oxygen content changes from 11.4% to 16.2% in the "base fluid used EN1-1" as well as from 11.4% to 13.1% in the fluid "with additives used EN1-1". Thus, the oxygen content significantly increases with usage in the hydraulic test bench under thermic-oxidative operating conditions.

Table 3b: ThOD and ThIC of the fluids calculated via elemental analysis (CHN-Analyzer)

	C (%)	H (%)	O (%) [a]	ThOD (g O_2/g)	ThIC (g CO_2/g)
gear fluids					
base fluid	76.1	12.5	11.4	2.92	2.79
base fluid blank EN1-0	75.8	11.1	13.1	2.78	2.78
base fluid used EN1-1	72.8	11.0	16.2	2.66	2.67
with additives	76.2	12.4	11.4	2.91	2.79
with additives used EN1-1	75.2	11.7	13.1	2.81	2.76
with additives used AGL-1	75.9	11.6	12.5	2.83	2.78
with additives used AGL-2	75.4	11.6	13.0	2.81	2.76

[a] O (%) = 100 – (H (%) + C (%)); ZDR = machine tool: cutting, EN1 = hydraulic test bench, -0 = blank, -1 = open system, AGL = plain bearing test bench, -1 = 80 °C, -2 = 90 °C

Biodegradability

The natural standard soil Lufa 2.2 is used as inoculum in the described biodegradation test system. For ready biodegradability test systems in water according to OECD guideline 301 the concentration of aerobic and facultatively anaerobic bacteria and fungi should be in the range of 10^6 to 10^8 cfu/l. Besides this, the oxygen consumption in the blanks should not exceed 10 mg/L within 28 days (10% of the ThOD). For this reason 2 g soil Lufa 2.2/l are used as inoculum in this work for the aquatic test system. Other soil samples with similar characteristics can also be used.

Biodegradation kinetics of the base fluids and the reference substance sodium benzoate are shown in *Figure 1*. The less distinct first order kinetic of the hydraulic base fluid can be explained by solubility and distribution characteristics of the substance, which causes transport limitations and may cause linear degradation kinetics. In these cases, the maximum rate of biodegradation is limited and a pass within the 10-day window is unlikely, because the saturation constant of the bacteria is high in relation to solubility. So this criterion for „ready" biodegradability is inappropriate, and in many cases incubation may need to be continued beyond 28 days before biodegradation reaches a plateau (Battersby, 2000). Therefore, the 10-day window criterion is not applied in the O_2/CO_2 Headspace Test with GC-TCD and only the degradability after 28 days is applied under qualitative consideration of the kinetics. In contrast to the hydraulic base fluid, the cutting base fluid shows an increase of biodegradation velocity with a similar lag-phase. The 10-day window criterion is passed, although water solubility is low.

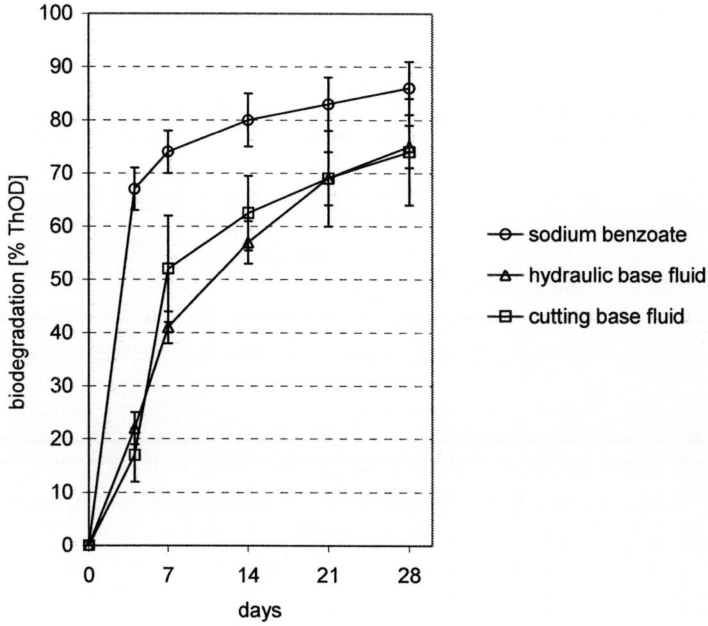

Figure 1: Biodegradation kinetics in water of sodium benzoate, hydraulic fluids and cutting base fluids; standard deviation of 3 parallels (Eisenträger et al. 2002)

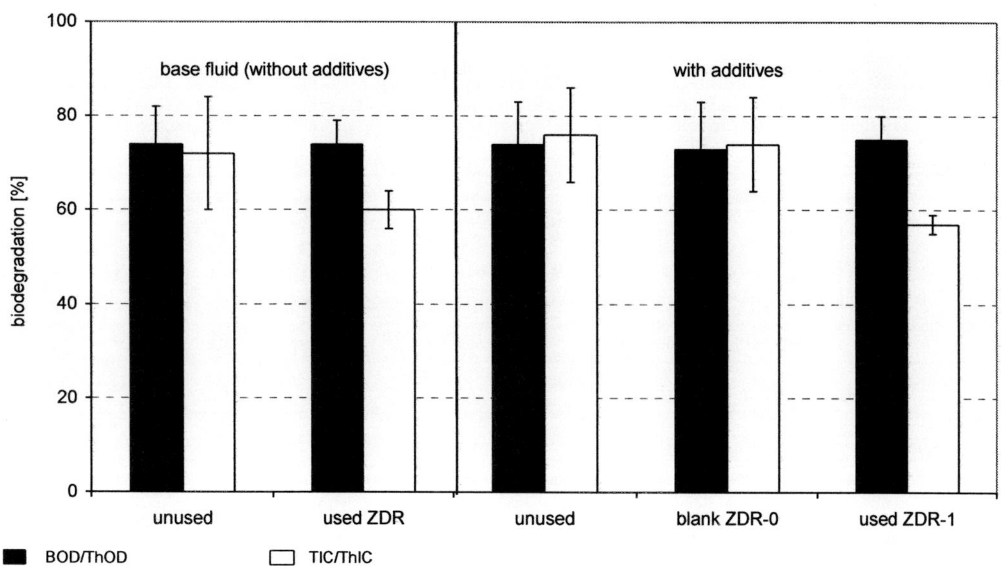

Figure 2: Biodegradability in water of cutting fluids; ZDR = machine tool with cutting operations; standard deviation of 3 parallels (Eisenträger et al. 2002)

During the cutting process the lubricants are exposed to high temperatures and pressures. Despite of these extreme conditions, no effects on biodegradability can be detected for the "base fluid used ZDR" as well as for the formulated fluid "with additives used ZDR-1" (*Figure 2*). This can be explained both by the large oil volume of 200 liter in this machine tool and by evaporation of used lubricants during operation. Therefore, only slight changes occur. Furthermore, no effects of additives on biodegradation can be observed.

The additives also have no respectively a slight influence on the biodegradability of the hydraulic fluids (*Figure 3*). Biodegradability of the oxopolymerised "base fluid used EN1-1" is lower than the biodegradability of the "unused base fluid" and the "base fluid blank EN1-0". It even is below the OECD limit value of 60% for ready biodegradability. In contrast to this, no usage effects can be detected in the hydraulic fluids with additives.

As expected, mineral oils are far less biodegradable than the ester oils and their biodegradability is not affected significantly by usage.

Results of biodegradability testing in soil are presented in *Figure 4* (hydraulic fluids) and *Figure 6* (gear fluids). For the hydraulic fluid it is clearly demonstrated that usage in the open system allowing oxygen intake causes a decrease of biodegradability both with and without additives (samples EN1-1). The increase of biodegradability of the mineral oil after usage might be due to enhanced bioavailability of the ingredients. Biodegradability after usage of the hydraulic fluids with additives in the plain bearing test bench differs significantly depending on the temperature: after usage at 90°C (sample AGL-2) it is lower than after usage at 80°C (sample AGL-1). The results indicate that biodegradability in soil is affected stronger by usage than biodegradability in water.

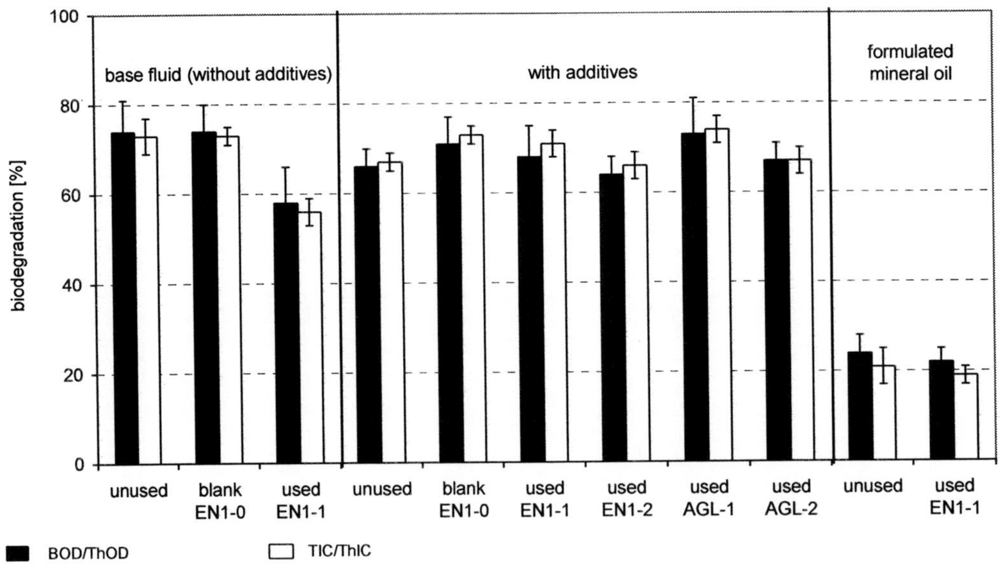

Figure 3: Biodegradability in water after 28 days of hydraulic fluids; EN1 = hydraulic test bench, -0 = blank, -1 = open system, -2 = closed system; AGL = plain bearing test bench, -1 = 80 °C, -2 = 90 °C; standard deviation of 3 parallels (Eisenträger et al. 2002)

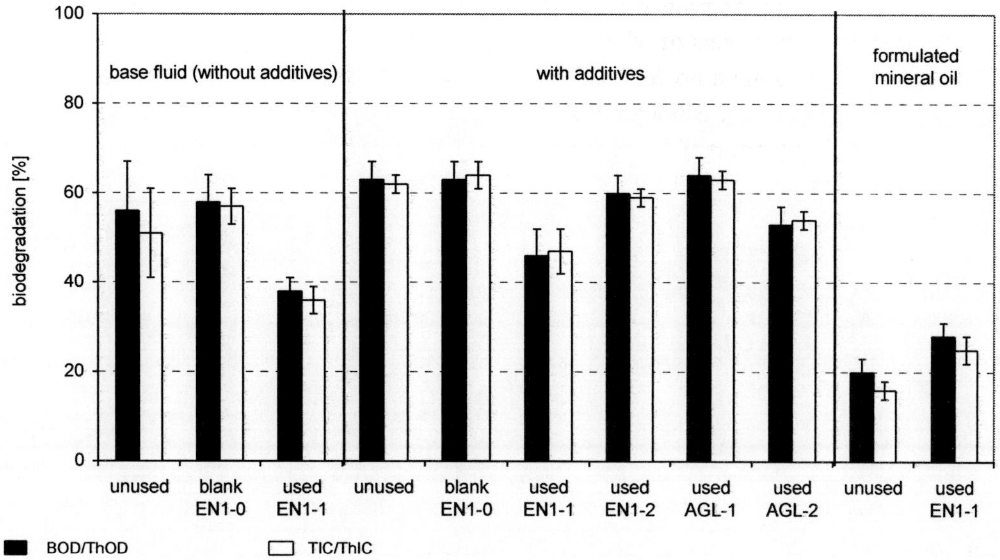

Figure 4: Biodegradability in soil after 32 days of hydraulic fluids; EN1 = hydraulic test bench, -0 = blank, -1 = open system, -2 = closed system; AGL = plain bearing test bench, -1 = 80 °C, -2 = 90 °C; standard deviation of 3 parallels

In contrast to the hydraulic fluid, the biodegradability of the gear fluid is not affected by usage (*Figures 5 and 6*). The viscosity of the gear fluid is higher than the one of the hydraulic fluid (*Table 1b*) corresponding to a lower biodegradability of the gear fluid both in soils and water. The strong increase of viscosity caused by usage does not result in a change of biodegradability.

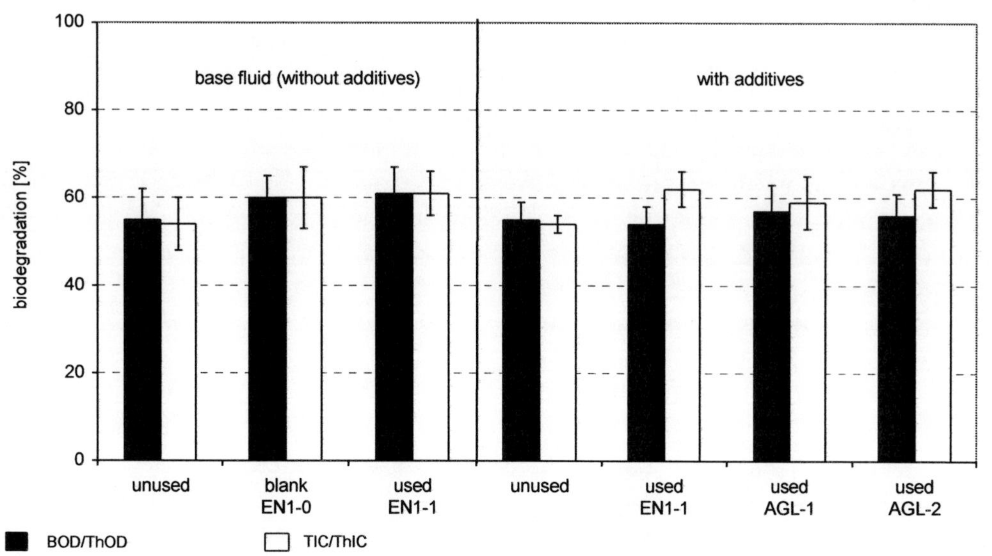

Figure 5. Biodegradability in water after 28 days of gear fluids; EN1 = hydraulic test bench, -0 = blank, -1 = open system, -2 = closed system; AGL = plain bearing test bench, -1 = 80 °C, -2 = 90 °C; standard deviation of 3 parallels

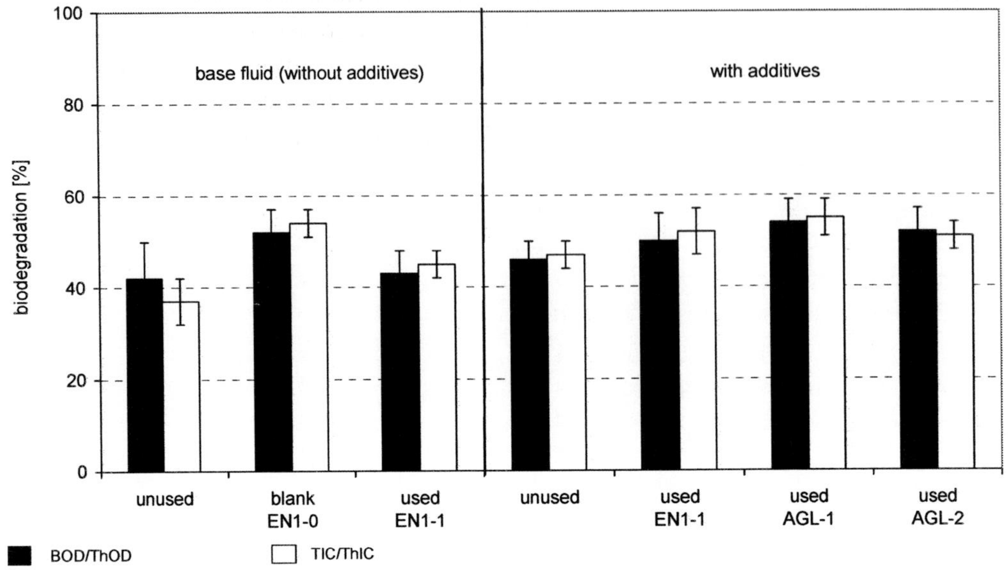

Figure 6. Biodegradability in soil after 32 days of gear fluids; EN1 = hydraulic test bench, -0 = blank, -1 = open system, -2 = closed system; AGL = plain bearing test bench, -1 = 80 °C, -2 = 90 °C; standard deviation of 3 parallels

DISCUSSION

Most oleochemical esters as base fluids fulfill the requirements of the eco-labels. However, the additives necessary to achieve the required technical performance are often the limiting components (Willing, 1999). Randles (1992) showed, that additives can have a significant positive or negative effect on biodegradation of esters, applying the CEC-test. For the fluids presented here, the influence of additives on biodegradability is not significant. As described by Battersby (2000), the biodegradability only depends on the kind of the base fluid. In contrast to the biodegradability, Bennett et al. (1990) demonstrated with mineral oil that ecotoxicity only depends on the additives and not on the kind of the base fluid. Thereby water accommodated fractions are applied for toxicity testing. Furthermore, the theoretical values of O_2 uptake and CO_2 production proposed by Battersby (2000) are confirmed by the results obtained with the mineral oils (ThOD = 3.4, ThIC = 3.1).

In this work, the fluids were tested under extreme operating conditions in machine tools and test benches. e.g. the "hydraulic fluid" is exposed to high temperature (90 – 120 °C), pressure (35 MPa) and shearing forces. For the investigation of thermic-oxidative aging processes this test bench can be operated openly or closed against oxygen. Despite of these experimental conditions no effects can be detected in water except for one sample without additives: biodegradability in this case was below 60% within 28 days. As the increase of viscosity and the CHN-analysis has been proved, the fluid was polymerised oxidatively. Such polymerisations could also be observed for hydraulic oils based on rapeseed oils (Zeman, 1997). Therefore, oxidative inhibitors are necessary as additives to prevent these changes. In contrast, Remmele and Widmann (1998) assumed, that there is no explicit relation between polymerisation with branched carbon chains and lower biodegradability.

Used lubricants may enter the aquatic or terrestrial environment via air or direct exposure. Therefore, biodegradability should be quantified both in water and soil. By comparing the results for both lubricants (Figures 3 and 4 for the hydraulic fluid and Figures 5 and 6 for the gear fluid), it can be stated that the biodegradability in soil after 32 days is slightly lower than in water after 28 days and that the effects of additives and usage are similar except for one testing condition: Biodegradability of the hydraulic fluid with additives after usage in the hydraulic test bench with oxygen intake is significantly lower in soil than in water. This might be caused by a change in bioavailabilty of some ingredients.

Necessary antioxidant concentrations can be reduced by changing the base fluid esters chemically without reducing performance characteristics of the oils. By these means, risks caused by toxic effects can be reduced, too. A promising approach is followed at the "Institute of Technical Chemistry and Heterogene Catalyses" at the RWTH Aachen. The double bindings of the fatty acids responsible for polymerisation are saturated with carbon acids. This results in an increase of oxidation stability and nearly unchanged filterability at low temperatures (Keller et al., 2000). This approach has already been cited by Wagner et al. (2001).

CONCLUSION

The combined batch assay "O_2/CO_2-Headspace Test with GC-TCD" is well suited for the testing of biodegradability of lubricants in water and soil. In this work, the effects of additives and usage are demonstrated. The additives in usual concentrations have no significant effect on the biodegradation of synthetic and native ester lubricants. This means that biodegradability mainly depends on the characteristics of the base fluids. Nevertheless, polymerisation effects may occur and reduce biodegradability if the specific additives preventing these effects are missing or if their concentration is lowered during usage. Further effects influencing biodegradability have not been observed in this work.

For these reasons, antioxidants like phenolic or aminic substances are necessary to prevent oxopolymerisation. These substances are often environmentally harmful and ecotoxic, and their concentration should be lowered by a simultaneous increase of oxidation stability of the base fluids.

Since the chemical characteristics of these and all other ingredients of the lubricants may differ extremely, biodegradability should be assessed and optimized both in soil and water. The main advantage of the "O_2/CO_2-Headspace Test with GC-TCD" besides quantification of both CO_2 and O_2 is, that biodegradability in soil and water can be assessed using a very similar method.

ACKNOWLEDGEMENTS

This work was supported by the Deutsche Forschungsgemeinschaft (DFG) (SFB 442 "Environmentally Friendly Tribosystems by suitable coatings and fluids with respect to the machine tool"; project "Methods and concepts for the risk assessment of tribological systems and machine tools referring to ecology and environmental medicine").

The cutting processes were kindly performed by the "Laboratory for Machine Tools and Production Engineering", the plain bearings by the "Institut for Machine Elements" and the elemental analysis (C, H, N) by the "Institute for organic chemistry" of the RWTH Aachen. The authors are grateful to M. Engels and S. Datta for technical assistance.

REFERENCES

Anonymous, 2003. Proposal for a regulation of the European Parliament and of the Council concerning the Registration, Evaluation, Authorisation and Restriction of Chemicals (REACH), establishing a European Chemicals Agency and amending Directive 1999/45/EC and Regulation (EC) {on Persistent Organic Pollutants} Proposal for a directive of the European Parliament and of the Council amending Council Directive 67/548/EEC in order to adapt it to Regulation (EC) of the European Parliament and of the Council concerning the registration, evaluation, authorisation and restriction of chemicals. *http://europa.eu.int/eur-lex/en/com/pdf/2003/com2003_0644en.html*

Bartz, W. J., 1998. Lubricants and the environment. *Tribology International* 31 (1-3), 35-47.

Battersby, N. S., 2000. The biodegradability and microbial toxicity testing of lubricants – some recommendations. *Chemosphere* 41 (7), 1011-1027.

Bennett, D., Girling, A. E., Bounds, A., 1990. Ecotoxicology of oil products: Preparation and characterization of aqueous test media. *Chemosphere* 21 (4-5), 659-669.

CEC L-33-A-93, 1995. *Biodegradability of two-stroke cycle outboard engine oils in water.* Co-ordinating European Council for the development of performance tests for lubricants and engine fuels method.

DIN 51558, 1979. Bestimmung der Neutralisationszahl (Measurement of total acid number). German Standard (Deutsches Institut für Normung), *Beuth Verlag*, Berlin, Germany.

DIN 51562, 1978. Messung der kinematischen Viskositaet mit dem Ubbelohde-Viskosimeter (Measurement of the kinematic viscosity with the Ubbelohde-viscosimeter). German Standard (Deutsches Institut für Normung), *Beuth Verlag*, Berlin, Germany.

Eisentraeger A., Schmidt M., Murrenhoff H., Dott W., Hahn S., 2002. Biodegradability testing of synthetic ester lubricants – effects of additives and usage. *Chemosphere* 48, 89-96. Figures and tables reprinted from *Chemosphere*, Vol. 48, Eisentraeger et al. Biodegradability testing of synthetic ester lubricants – effects of additives and usage, pages no. 89-96, copyright 2002 with permission from Elsevier.

Haigh, S. D., 1995. Fate and effects of synthetic lubricants in soil: biodegradation and effect on crops in field studies. *The Science of Total Environment* 168, 71-83.

Hita, C., Parlanti, E., Jambu, P., Joffre, J., Amblèsa., 1996. Triglyceride degradation in soil. *Organic Geochemistry* 25 (1/2), 19-28.

ISO 11266, 1994. *Soil quality - Guidance on laboratory testing for biodegradation of organic chemicals in soil under aerobic conditions.* International Standard.

ISO/TR 15462, 1997. Water quality – Selection of tests for biodegradability. International Technical Report, *International Organization of Standardization*, Genève, Switzerland.

Keller, U., Fischer, J., Hoelderich, W. F. et al., 2000. Neue Schmierstoffe auf der Basis nachwachsender Rohstoffe: Oekotoxikologische und oxidative Eigenschaften (New lubricants from renewable resources: Ecotoxic and oxidative characteristics). *O+P Oelhydraulik und Pneumatik* 4, 240-245.

OECD, 1992. Guidelines for testing of chemicals: Summary of considerations in the report from the OECD expert group on degradation and accumulation. *Organization for Economic Co-operation and Development*, Paris.

Pagga, U., 1997. Testing biodegradability with standardized methods. *Chemosphere* 35 (12), 2953-2972.

RAL-UZ 79, 1996. *Grundlage für Umweltzeichenvergabe: Biologisch schnell abbaubare Hydraulikfluessigkeiten (Basics of eco-labelling: ready biodegradable hydraulic fluids).* Deutsches Institut für Gütesicherung und Kennzeichnung e.V. (RAL), Umweltbundesamt, Jury Umweltzeichen.

Randles, S. J., 1992. Environmentally considerate ester lubricants for the automotive and engineering industries. *Journal of Synthetic Lubrication* 9, 145-161.

Remmele, E., Widmann, B., 1998. Hydraulic fluids based on rapeseed oil in agricultural machinery – Suitability and environmental impact during use. In: Bartz, W. J. (Eds.). Industrial and Automotive Lubrication Vol.I. 11th International Colloquium 13-15 January 1998, *TA Esslingen*, Germany.

Schuelert, G., Bernhard, U., 1995. Alterungsverhalten und Veränderungen der biologischen Abbaubarkeit umweltschonender Hydraulikfluessigkeiten (Aging behavior and changes of biodegradability environmental gently hydraulic fluids). Conference of the Technical Academy of Esslingen „Biologisch schnell abbaubare Schmierstoffe und Arbeitsflüssigkeiten, speziell auf pflanzlicher Basis (Ready biodegradable lubricants and working fluids, especially on native basis)". *Esslingen*, February 1995.

Wagner, H., Luther, R., Mang, T., 2001. Lubricant base fluids based on renewable raw materials - Their catalytic manufacture and modification. *Applied Catalyst A: General* 221, 429-442.

Willing, A., 1999. Oleochemical esters – environmentally compatible raw materials for oils and lubricants from renewable resources. *Fett/Lipid* 101 (6), 192-198.

Willing, A., 2001. Lubricants based on renewable resources – an environmentally compatible alternative to mineral oil products. *Chemosphere* 43, 89-98.

Zeman, A., 1997. Biologisch abbaubare Kuehlschmierstoffe und Hydraulikfluessigkeiten (Biodegradable metal working and hydraulic fluids). *Tribologie und Schmierungstechnik* 44 (4), 160-165.

In: Water Pollution: New Research
Editor: A. R. Burk, pp. 161-180
ISBN: 1-59454-393-3

Chapter 8

REMOVAL OF HALOGENATED HYDROCARBONS FROM WATER BY ZERO-VALENT IRON

Vaclav Janda[1] ***and Ondra Sracek***[2]
[1]Department of Water Technology and Environmental Engineering,
Institute of Chemical Technology, Prague, Czech Republic
[2]Institute of Geological Sciences, Faculty of Science,
Masaryk University, Brno, Czech Republic

ABSTRACT

Halogenated organic compounds in water can react under anaerobic conditions with zero-valent iron Fe^0. While halogenated compounds are reduced via intermediates to non-toxic simple hydrocarbons, Fe^0 is oxidized to Fe^{2+}. An excess of hydrogen gas is also produced by the reaction of iron with water. Dehalogenation reaction can be utilized for decontamination of underground water in passive permeable reactive barriers, where water flows spontaneously through a layer of granular iron. The basic parameters of the dehalogenation reaction and field application of the reaction including applications of isotopes and geochemical modeling are discussed in this chapter.

INTRODUCTION

Halogenated hydrocarbons present some of the most persistent environmental contaminants and are of concern worldwide. Traditional disposal practices of chlorinated compounds such as trichloroethene (TCE) and tetrachloroethene (PCE) contribute to groundwater contamination and industrial wastewater pollution all over the world, whereas TCE and PCE were widely used mainly for dry cleaning and metal surface degreasing. Technologies used for the removal of these compounds depend principally on their

[1] E-mail: vaclav.janda@vscht.cz
[2] e-mail: srondra@yahoo.com

concentration in water and on the subsequent usage of the treated water. Processes based on sorption (like sorption on granulated activated carbon), gas stripping and biological methods are the most frequently used in water treatment practice. All the above mentioned processes exhibit certain, and in some cases principal, disadvantages. Sorption materials are expensive and their lifetime is limited. Additionally, there is a problem with their save disposal. Stripping processes remove the toxic compounds from water but transport them into air, which results in atmospheric pollution. For halogenated hydrocarbons, biological processes are efficient enough only in an anaerobic environment and are rather slow. Decontamination of polluted groundwater is always problematic since technological facilities used for flow-through treatment system must be designed for a high capacity, in order to clean usually huge amounts of groundwater in aquifer in an acceptable period of time. Therefore, the economic considerations and the concern about effective long-term treatment are contributing to a shift away from historic flow-through (pump and treat) treatment remediation systems towards in-situ methods.

DEHALOGENATION BY ZERO-VALENT IRON – BASIC DATA

Metallic iron provides an alternative opportunity to degrade chlorinated organics abiotically. Under anaerobic conditions, zero valent iron (ZVI) is able to reduce chlorinated hydrocarbons via intermediates, to non-toxic hydrocarbons. The reaction does not require elevated temperature. The reaction is thermodynamically feasible and is limited kinetically. There are probably two ways of chlorinated hydrocarbons reduction (Arnold and Roberts, 2000a):

$$\text{Hydrogenolysis: } RX + 2e^- + H^+ = RH + X^- \tag{1}$$

For example, tetrachloromethane is reduced via trichloromethane to dichloromethane. The lower the degree of chlorination the slower the rate of dechlorination reaction (Matheson and Tratnyek, 1994; Johnson et al., 1998).

$$\text{Reductive elimination: } \text{-CHX-CHX-} + 2e^- = \text{-CH=CH-} + 2X^- \tag{2}$$

In order for this reaction to take place, the halogen atoms must be in 1,2-positions.

It seems that reaction 2. prevails during reduction of chlorinated ethenes, therefore, reaction intermediate products of reaction 1. (dichloroethene and vinylchloride) are never present in water in high concentrations. This fact is of great importance because e.g. vinylchloride exhibits higher toxicity than the parent compounds.

ZVI serves as donor of electrons (reducing agent) for the reductive dechlorination and hydrogenolysis. Under anaerobic conditions (redox potential can drop to -500 mV) the elemental iron is oxidized to ferrous ion,

$$Fe = Fe^{2+} + 2e^- \tag{3}$$

which subsequently hydrolyzes to ferrous hydroxide.

$$Fe^{2+} + 2H_2O = Fe(OH)_2 + 2H^+ \quad (4)$$

Simultaneously, hydrogen gas is formed by reaction of iron with water.

$$Fe + 2H_2O = Fe(OH)_2 + H_2 \quad (5)$$

Final products of the reaction are simple hydrocarbons like methane, ethene, ethane and acetylene (Schreier and Reinhard, 1994; Orth and Gillham, 1996; Roberts et al., 1996; Allen-King et al., 1997; Campbell et al., 1997; Su and Puls, 1999; Farrell et. al, 2000). ZVI iron also reacts with water producing hydrogen gas and hydroxide ions (pH of water increases). The most prevailing compound in the gas phase is hydrogen. Substantial part of the dehalogenation reaction products is methane (in case of single carbon hydrocarbons like tri- or tetrachloromethane) or ethene with ethane for chlorinated ethanes and ethenes (see Figs. 1 and 2 – batch experiments, volume of reaction vessel 22 mL, 5 g of iron, grain size of iron 150 μm, specific surface area of iron 0.66 m^2/g, temperature 22 °C). Other hydrocarbons than those shown in Figs. 1 and 2 are present in negligible concentrations.

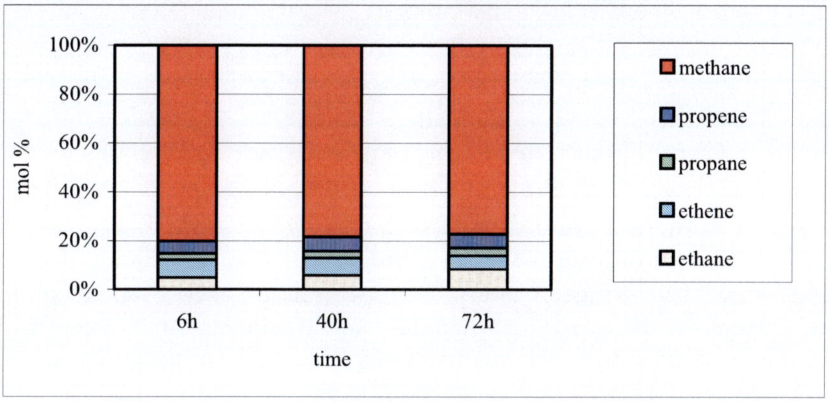

Fig. 1. Distribution of gaseous products resulting from the dehalogenation of trichloromethane in water as a function of reaction time – batch experiment. Adapted from Vasek (2002).

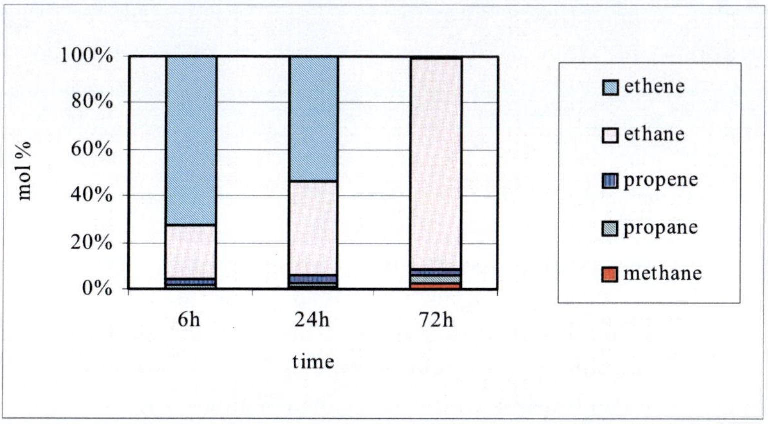

Fig. 2. Distribution of gaseous products resulting from the dehalogenation of trichloroethene in water as a function of reaction time – batch experiment. Adapted from Vasek (2002).

It is assumed (and partly confirmed experimentally) that the bulk dehalogenation reaction can be roughly described by the first order equation (Burris et al., 1995; Johnson et al., 1996; Wüst et al., 1999; Deng et al. 1999; Weerasooiriya and Dharmasena, 2001). Some authors suggested combination of the 1st and the 2nd order equation (Wüst et al., 1999) with respect to concentration of halogenated hydrocarbon in water. A two-phase kinetic model has been also developed for the general class of reactions involving reaction in one phase and mass transfer between two phases (Gotpagar et al., 1998). The dehalogenation reactions were also modeled by power law models, pseudo-stationary models with a controlling surface reaction rate and non-stationary models without the assumption of rate controlling step. It was found that power law models, Langmuir–Hinshelwood (or Langmuir-Hinshelwood-Hougen-Watson) analogy models and general models of heterogeneous reactions are reliable for the kinetic description of dechlorination (Arnold and Roberts, 2000a; Arnold and Roberts, 2000b; Janda et al., 2004) – see Fig. 3: reaction conditions the same as in the previous paragraph. Dechlorination of mixtures of chlorinated hydrocarbons with zero-valent metals provided information concerning the kinetics of simultaneous degradation of different contaminants. It was found that competitive effects (loading of the reactive iron surface by a single contaminant) as well as interspecies competition of several contaminants for the reactive sites available can take place (Schäfer, 2003). Since the reaction is heterogenous, its rate is also proportional to a specific area of the iron used. The smaller is the particle size of iron, the higher is the specific surface area. So called nanoscale Fe^0 (particle size 1–200 nm), characterized by high specific surface area and high reactivity, rapidly transforms chlorinated hydrocarbons under ambient conditions, which results in complete disappearance of the parent compounds from the aqueous phase within a few minutes (Lien and Zhang, 1999; Choe et al., 2001; Lien and Zhang, 2001). It is not exactly known, which process is limiting: adsorption on iron surface, diffusion and/or chemical reaction itself. Activation energy of the reaction varies from 30 to 80 kJ/mol (Schreier and Reinhard, 1997; Deng et al., 1999), according to the particular chlorinated hydrocarbon. Usage of surfactants to increase the concentrations of halogenated hydrocarbons near the iron surface by sorption into surfactant layers was insufficient to increase reduction (Lorain, 2001).

As stated above, during the dehalogenation reaction and reduction of water by ZVI, the pH of water solution increases. In anaerobic conditions $Fe(OH)_2$ precipitates and finally the reaction takes place in saturated solution of $Fe(OH)_2$. The degradation of TCE was not observed at pH values of 9 and 10 (Slater et al., 2002). Decreasing of the solution pH increases the degradation rate. Some authors (Weerasooriya and Dharmasena, 2001) suggest addition of pyrite to iron as a buffer. Due to presence of disulfidic group in the pyrite, the pH value can be kept within the 6.7 – 7.8 range. Increase of pH strongly depends also on concentration of chlorinated hydrocarbons in water and the natural neutralization capacity of water (HCO_3^-- CO_2 buffering system). Precipitated layer can prevent contact of aqueous and solid iron phase and block the iron surface for the reaction with chlorinated hydrocarbons (Johnson et al., 1998; Odziemkowski, 1998), which has in field applications serious consequences (see paragraph 3.4. of this chapter). This phenomenon refers also to naturally present oxides coverage on the iron. Therefore, some authors investigated iron pre-treatment, e.g. by hydrochloric acid or ultrasound to enhance the dehalogenation reaction (Geiger et al., 2002). If the natural water system is buffered weakly, the pH value can increase above 10 (Liang et al., 2000).

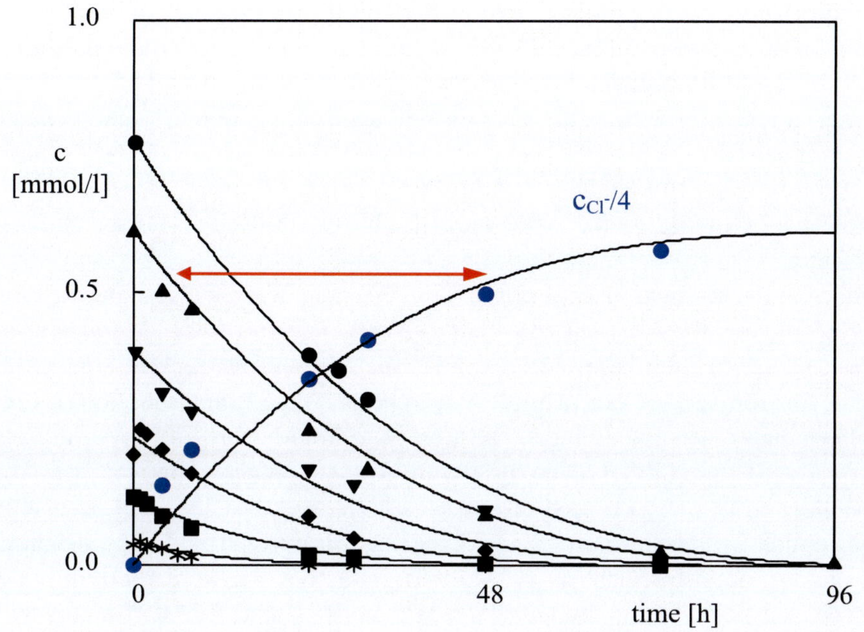

Fig. 3 Removal of tetrachloroethene from water for its different initial concentrations and increase of chloride ion concentration due to a change of organic chlorine to chloride (only one curve for one concentration of tetrachloroethene is shown – see arrow). Adapted from Janda et al. (2004).

In real systems, the formation of precipitates depends on composition of the contaminated water. As shown above, the first product is $Fe(OH)_2$ ($K_s = 8\cdot10^{-16}$) – Liang et al., 2000; Mackenzie et al., 1999. When higher concentrations of carbon dioxide, bicarbonates, calcium and eventually carbonates are present, formation of $FeCO_3$ ($K_s = 3{,}2\cdot10^{-11}$) and $CaCO_3$ ($K_s = 2{,}8\cdot10^{-9}$) can be expected (Mackenzie et al., 1999; Liang et al., 2000). If hydrogen sulfide occurs in groundwater, precipitation of FeS ($K_s = 1{,}59\cdot10^{-19}$) could take place (Mackenzie et al., 1999). Hydrogen sulfide can arise from sulfate reduction, when redox potential is lower than – 50 mV. $Fe(OH)_2$ is thermodynamically unstable and in the presence of dissolved oxygen is oxidized to $Fe(OH)_3$ ($K_s = 2{,}64\cdot10^{-39}$) and/or Fe_3O_4 ($K_s = 1{,}70\cdot10^{-52}$) at pH higher than 6 – 7. The presence of dissolved oxygen can also cause passivation of the ZVI surface through (hydro)oxides formation. This thin layer prevents transfer of electrons and decreases dehalogenation reactions. In paper by McGuire et al. (2003) the utility of surface analysis techniques, including electron spectroscopies, vibrational spectroscopies, and atomic force microscopy in elucidating the roles played by the surface was illustrated. Moreover, hydrogen gas can form a thin film onto the metallic iron surface, which disables contact of the chlorinated hydrocarbons and active surface of the iron.

Besides precipitation reactions, there are other side reactions affecting water composition after its treatment. For example, ZVI reduces nitrates completely to ammonia (Cheng et al., 1997):

$$NO_3^- + 10\,H^+ + 4\,Fe^0 = NH_4^+ + 3\,H_2O + 4\,Fe^{2+} \tag{6}$$

This reaction may cause problems owing to a high concentration of ammonia in final water (1 mg of nitrates produces 0.29 mg of ammonia). It also contributes to a great concentration of ferrous ions in water and increasing of pH value of water.

Decontamination of waters using ZVI is not applicable only to volatile halogenated hydrocarbons. It can also be used for the removal of DDT (Sayles et al., 1997), some other pesticides (Monson et al, 1998; Ghauch et al, 2001; Gibb et al., 2004; Dombek et al., 2004), residual explosives (Hundal et al., 1997) and metals (Cantrell et al., 1995; Blowes et al., 2000) from water.

The reductive dechlorination does not take place exclusively with elemental iron. Other zero-valent metals, e.g. Cu/Al (Lien and Zhang, 2002), Sn, Zn or Mg (Morales et al.,2002; Boronina et al., 1995) can be used as well. Microorganisms present under anaerobic conditions on the iron surface can play also an important role (Lampron, 2001). Catalytical effect of some metals (Lien and Zhang, 2001; Centi and Rarathoner, 2003) was also observed. Presence of sulfur (Hasan, 2000) in mixture with iron can enhance the reaction. A reaction mechanism based on in-situ formation of ferrous sulfide as the catalyst and its reaction with the halocompounds in the presence of hydrogen has been postulated. The dehalogenation reaction need not be performed only by ZVI: reductive dechlorination of chlorinated methanes in cement slurries containing Fe(II) (Hwang and Batchelor, 2002) and/or Fe-bearing clay minerals (Cervini-Silva, 2002) were also described. Degradation kinetics of carbon tetrachloride and chloroform were generally very rapid with reaction rates comparable to rates that can be obtained with ZVI.

It is quite interesting that during reductive dechlorination with iron the reaction produces a systematic enrichment of ^{13}C isotope in the remaining substrate (Schüth et al., 2003; Dayan et al., 1999; Slater et al., 2002). This fact can serve as a tool for monitoring pilot-plant or full-scale technological applications performance, e.g. to distinguish between insufficient residence time for the reaction to take place and possible saturation of the reaction zone by a contaminant (see paragraph 3.6. of this chapter). In the latter case the isotopic ratio should remain less unchanged than in the former one.

FIELD APPLICATIONS

Types of and Permeable Reactive Barriers (PRB) Choice of Material

Field applications of (PRB) are relatively recent issue. In 1970's and 1980's, principal remediation approach was pumping and treatment of contaminated ground water. It has been found that this approach generally did not work for complete remediation of contaminated sites. There were several reasons of this failure: (1) heterogeneity of geological environment, where ground water is pumped preferentially from permeable zones and transfer of mass from low permeability zones is mostly by diffusion (Mackay and Cherry, 1989), (2) presence of free phase, which dissolves slowly and is gradually enriched in low solubility compounds (Pankow and Cherry, 1996), and (3) non-linear and sometimes kinetically constrained desorption of contaminants (Appelo and Postma, 1996). The situation resulted in development of new remediation approaches like application of PRB.

Probably the first known application of PRB is the use of ZVI for degradation of pesticides (Sweeney and Fischer, 1972). However, very fast boom of this approach is related to research of R.W. Gillham's group on application of ZVI in PRB for degradation of chlorinated solvents and D.W. Blowes' group on application of PRB for remediation of inorganic contaminants like metals. Both groups are from the University of Waterloo in Canada. Currently (in early 2004) the number of PRB operating in the North America is more than 40, and there are also several PRB applications in Europe.

In this text our focus is on degradation of chlorinated solvents by the application of ZVI, but the same approach is also used, for example, for remediation of Cr(VI). From viewpoint of PRB geometry, two basic types of barriers can be recognized (Gavaskar et al., 1998). They are a continuous barrier and so called funnel and gate system (Fig 4). A continuous PRB consists of a single reactive cell containing reactive material. A funnel and gate system has an impermeable section (funnel) which directs ground water flowing from PRB capture zone towards permeable zone with reactive material (gate). Both systems have their advantages and disadvantages. For example, at sites with very non-uniform distribution of contamination a funnel and gate system may better homogenize contaminant concentration before entering the reactive zone.

There is no universal type of iron, which could be used as reactive material in PRB. Generally it is crude iron with valence zero produced in car industry. This material should be over 95 wt % of Fe(0), free of grease, with low content of carbon, and without surface coatings caused by pre-oxidation. In some cases, presence of traces of pyrite was found to be beneficial because its oxidation keeps pH lower and, thus, prevents precipitation of large amounts of secondary minerals in the barrier (see later).

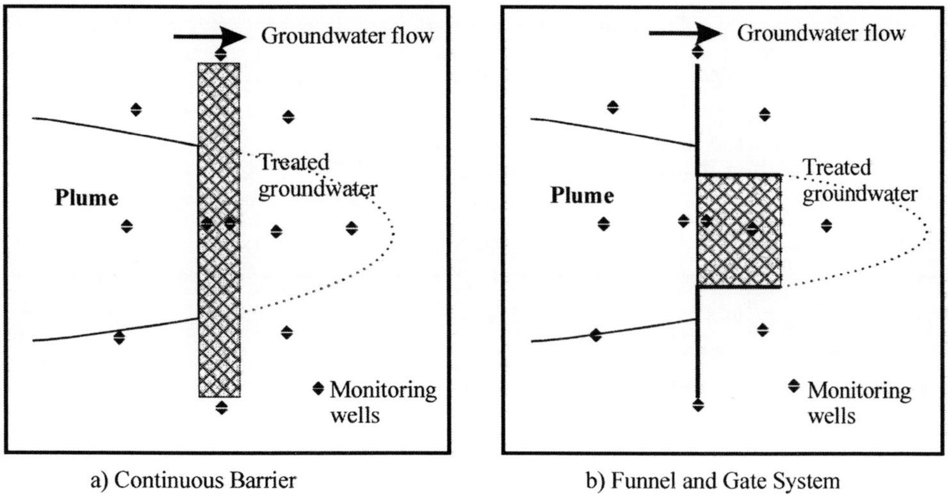

Fig. 4 Basic types of barriers.

Determination of Rate Constants and Dimensions of PRB

The sequence of material testing follows the order batch test → column test → pilot test (Gillham and O'Hannesin, 1994). Principal objectives of preliminary testing to select a suitable reactive material and also to determine rate constant. As we stated earlier, assumption about 1st order of reaction is a gross simplification and reaction pathway is more complicated. However, in many situations first order kinetics can reproduce measured data and, thus, can be used for design of PRB.

Batch test is performed in closed recipient without flow of water. In that case, reactive material is located in a flask filled by water contaminated by chlorinated solvents. Concentrations of chlorinated solvents are plotted as a function of time. When points in graph $\ln(C/C_0) - t$ form a straight line, then the reaction is considered to be of pseudo-1st order (pseudo- because the influence of other factors is considered constant). Resulting rate constant k [$time^{-1}$] is obtained as slope of the graph. Results of this type of test generally have a limited applicability because values of rate constant are underestimated. This is caused by limited contact of contaminated water with ZVI compared to conditions in PRB.

Column test experiments are performed to obtain more representative values of rate constant. In that case conditions are dynamic and velocity of water flow in a column is kept at value similar to field conditions. It is convenient to use ground water from contaminated field site because other geochemical reactions in PRB can be evaluated (see later). Sampling ports in the column should be located not only at the inflow and outflow points, but also in the reactive material. After flushing of several pore volumes, steady-state in the column is established. Concentrations as a function of distance x from the beginning of the column are converted to concentrations as a function of time using known velocity v of water flow in the column. Then rate constant is obtained as a slope of graph $\ln(C/C_0) - t$ just like in the case of batch test. Water flow velocity can be obtained from tracer test with a conservative tracer (for example, Cl^-) or calculated from known volumetric flux, cross section of a column, and effective porosity of reactive material. Chloride mass balance calculations can be performed to verify that degradation of chlorinated solvents really occurs (for example, Baciocchi et al., 2003). In the case of column tests, values of rate constant are generally more representative than values obtained from batch tests.

Rate constant depends on reactive surface, and, thus, on the degree of compaction of reactive material. When degree of compaction increases, so does the value of rate constant because more ZVI is in contact with contaminated water. In literature the rate constant is often normalized to m^2/mL to compare values for different specific surfaces. Specific surface of ZVI in m^2/g is determined by methods based on adsorption of a gas like BET, and other parameters necessary for conversion are specific mass and porosity of reactive material.

When value of rate constant from laboratory is used, several corrections have to be made for different field conditions. First of all, temperature of ground water is lower than temperature in laboratory and rate of reaction is slower. As a rough approximation, rate of reaction decreases by a factor of 2 for drop of temperature about 10°C (Gavaskar et al., 1998). More exact calculation is by the application of the Arrhenius plot for reaction kinetics. Rate of reaction is

$$k = A.\exp\frac{-E_a}{R.T} \tag{7}$$

where k is rate constant, A is pre-exponencial factor, E_a is activation energy (for dehalogenation reaction calculated e.g. by Schreier and Reinhard, 1994 or Deng et al., 1999), R is universal gas constant, and T is temperature. After taking of natural logarithms we obtain

$$\ln k = \ln A - \left(\frac{E_a}{R.T} \right) \tag{8}$$

When experiments at least two different temperatures are performed, the graph ln $k - 1/T$ with slope $-(E_a/R)$ and intercept ln A is plotted. Then the rate constant can be calculated from Equation 8 for any temperature. Other corrections are for lower degree of compaction in the field PRB. A common correction factor for rate constant determined in laboratory is 2.5 – 3.0 (Gavaskar et al., 1998).

A theoretical thickness of PRB is calculated using corrected rate constant, ground water flow velocity determined in the field, initial concentration, and remediation goal:

$$x_{PRB} = -\left(\frac{v}{k} \right). \ln\left(\frac{C_{limit}}{C_{initial}} \right) \tag{9}$$

where x_{PRB} is thickness of reactive barrier, v is water flow velocity within a PRB, k is rate constant, C_{limit} is remediation goal for the slowest decaying contaminant, and $C_{initial}$ is initial concentration. It is evident from the equation above that thickness of PRB increases when flow velocity is high and rate constant is low. This is contradictory to hydraulic requirement of higher permeability of PRB compared to surrounding aquifer because a contaminant plume would by-pass PRB in opposite case (see next section).

It is worth to mention that the equation (9) is applicable for remediation of other contaminants (hexavalent chromium etc.), when their removal can be approximated as 1st order kinetics process.

Hydraulic Aspects of PRB

Emplacement of reactive barrier into geological media represents a very serious disturbance of natural flow pattern. Principal reason is higher permeability of PRB compared to the permeability of aquifer. Continuous PRB disturbs flow pattern less than funnel and gate system. Two factors are very important for hydraulic design of PRB: capture zone of PRB and residence time of contaminated water in reactive zone of PRB.

Capture zone of PRB is space, from which all water particles will enter reactive zone of the PRB. In the case of continuous barrier the width of capture zone is larger than the width of the barrier. This is caused by refraction of flow lines towards more permeable barrier. This means that the barrier represents a preferential flow zone. Ratio between width of capture zone and width of PRB increases to a certain value of hydraulic conductivity and then remains constant. In the case of funnel and gate system the ratio depends on geometry of the system.

The contaminant plume has to be located completely within capture zone of PRB because some contaminated water could by-pass the barrier in an opposite case. This condition requires detailed monitoring of seasonal flow direction changes. These changes may occur during year as a consequence of varying recharge, changes of hydraulic function of surface streams at boundaries of an aquifer, and, first of all, due to anthropogennic factors like pumping of ground water. Under natural conditions flow direction may fluctuate in an interval of 30°, but there can but fluctuations within 180° interval in the proximity of surface water streams. Human induced ground water pumping may change water flow direction in an interval up to 90°. Regarding water flow changes, continuous barrier is more sensitive than funnel and gate system.

Porous media is generally very heterogeneous and advective transport of contaminants occurs principally in more permeable zones. This means that the barrier has to be designed for remediation of contaminated water from the most permeable zones. Other possibility is homogenization of flow in gravel zone located in front of reactive zone.

It is evident that quality and quantity of data necessary for hydraulic design of PRB is higher than in the case of more simple remediation systems. Integral part of hydraulic design became modeling of flow and transport for different configurations of barriers. Pioneering work considering various designs of funnel and gate systems was done by Starr and Cherry (1994). Common programs applied are MODFLOW (McDonald and Harbaugh, 1988) in combination with MODPATH (Pollock, 1989). These programs can visualize ground water flow using particle tracking approach. In that case trajectories of particles released in the proximity of PRB are traced. More complex models are represented by program FRAC3D (Therrien and Sudicky, 1995; Shikaze, 1996). It has to be emphasized that modeling of flow is a prerequisite for modeling of transport with geochemical reactions discussed later in this text. Principal results of hydraulic performance modeling of PRB can be summarized as follows:

- Maximum capture zone of PRB in funnel and gate system is achieved for apex angle (angle between two segments of funnel) equal to 180° (straight funnel). For any apex angle, the maximum discharge occurs for funnel perpendicular to ground water flow gradient.
- For funnel and gate system with funnel walls at 180° (straight funnel), the discharge through the gate and width of PRB capture zone increase when funnel width increases, but the increase is not directly proportional to the change of funnel width.
- For a given funnel and gate design, the discharge through a gate increases with increase of K_{gate} compared to $K_{aquifer}$. However, there is little change for $K_{gate}>10K_{aquifer}$. This means that K_{gate} does not have to be much higher than $K_{aquifer}$.
- There has to be a balance between maximizing of capture zone of PRB and maximizing residence time in the PRB. The residence time can be increased without affecting the capture zone by the increase of PRB width.

Precipitation of Secondary Minerals and Longevity of PRB

As discussed previously, precipitation of secondary minerals is of a very serious concern for prediction of PRB performance in the future. It is evident from the equation of anareobic corrosion that there generally is a very significant increase of pH within a PRB. This is followed by precipitation of secondary minerals such as calcite

$$Ca^{2+} + 2HCO_3^- = CaCO_3(s) + H_2CO_3 \quad (10)$$

siderite

$$Fe^{2+} + HCO_3^- = FeCO_3(s) + H^+ \quad (11)$$

and mackinawite

$$Fe^{2+} + HS^- = FeS(s) + H^+ \quad (12)$$

Mackinawite is a precursor of later formation of pyrite (Benning et al., 2000). This mineral phase is quite stable under low Eh and high pH conditions and rate of transformation to pyrite is very slow. Other mineral phases are various forms of so called Green Rust (carbonate Green Rust, $Fe_6(OH)_{12}CO_3$, and sulfate Green Rust, $Fe_6(OH)_{12}SO_4$), Bourrié et al. (1999), Génin et al. (2001). Fe(II) is formed in PRB in the process of anareobic corrosion, but bicarbonate and sulfate, which is converted by reduction to HS^-, are generally present in background ground water. This means that the higher are background inorganic carbon and sulfate concentrations, the higher generally also is the rate of precipitation of secondary minerals. Presence of nitrate in background ground water may have an adverse impact on degradation rates because nitrate interferes with removal of Fe_2O_3 layer formed on the surface ZVI in production of iron (Ritter et al., 2003). Under normal conditions, removal of the Fe_2O_3 layer is fast (Ritter et al., 2002). Hydrogeochemical data from two long-term monitored sites, from Elizabeth City (EC) in South Carolina and from Denver Federal Center (DFC) are in Table I.

Table I. Hydrogeochemical parameters of water at two PRB sites (adapted from Wilkin et al., 2003), EC-Elizabeth City, DFC-Denver Federal Center (Gate 1)

Sample	pH	E_h [mV]	O_2 [mg/l]	Ca [mg/l]	Fe [mg/l]	SO_4 [mg/l]	NO_3 [mg/l]	DIC [mg/l]	TDS [mg/l]
EC-entry	5.84	491	0.3	16.4	< 0.04	49	0.9	15	290
EC-PRB	9.51	-347	0.1	5.3	0.05	<1	< 0.1	8	243
DFC-entry	7.24	233	0.9	100	0.19	234	3.2	86	1090
DFC-PRB	10.30	-247	0.1	1.2	< 0.04	<1	< 0.1	60	510

Background ground water at EC site has much lower concentrations of sulfate and DIC than at DFC site. There is a very significant decrease in sulfate concentration in PRB at both sites. Also concentration of nitrate decreases below detection limit. Reduction of Ca and DIC concentrations are also significant. Regarding much higher background concentrations at DFC site, much higher precipitation rate can be expected. Calculated precipitation rates for inorganic C and S were 2.16 $kg.m^{-2}.year^{-1}$ and 0.80 $kg.m^{-2}.year^{-1}$ at EC site, but only 0.09 $kg.m^{-2}.year$ and 0.02 $kg.m^{-2}.year^{-1}$ at EC site (Wilkin et al., 2003).

Another site with relatively long-term monitoring record is Monkstown in Northern Ireland (Beck et al., 2001). At this site, concentrations of chlorinated solvents are very high (TCE up to 390 mg/L) and flow rate is relatively low due to low permeability of the aquifer. Precipitated crusts of calcite and siderite on the ZVI grains were found in narrow zone (5-10 cm) from the entry of PRB, but signs of corrosion of ZVI were found at the distance 40 cm from the entry of PRB. No signs of biological fouling were detected on the surface of ZVI. Estimated time before maintenance was between 10 and 15 years.

Generation of gases in PRB in redox reactions (for example, production of nitrogen by denitrification and hydrogen by anaerobic corrosion - Reardon, 1995) may also have an adverse impact on hydraulics and residence time, but the effect is considered minor compared to the effect of secondary minerals precipitation (Kamolpornwijit et al., 2003).

In summary, precipitation of secondary minerals is an important process, which is especially significant at the zone of contaminated water entry, e.g., in the first decimeters of reactive material at the inflow side. There are only limited data of long-term performance of PRB (Beck et al., 2001; Wilkin et al., 2003), but significant reduction of PRB permeability has not been found. However, more important seems to be coating of ZVI grains by secondary minerals, and, thus, reduction of surface available for reaction. Currently it seems that this is relatively long term process, but only limited long term monitoring data (less than 10 years) are available so far.

Data Acquisition for Design and Monitoring of PRB

Data for design of a reactive barrier can be divided broadly into ground water flow characteristics, organic composition of ground water, and inorganic composition of ground water (Gavaskar et al., 1998). Ground water flow characteristics include stratigraphy and soil types, depth of confining unit, depth of water table, ground water flow direction and its seasonal fluctuations, hydraulic conductivity, porosity, ground water flow velocity, and dimensions and depth of contaminant plume.

Organic composition of ground water includes spatial distribution of chlorinated solvents. Also processes affecting contaminant concentrations have to be known (dispersion, adsorption/retardation, biodegradation etc.).

Inorganic parameters include field parameters like pH, redox potential (Eh), and dissolved oxygen (DO). These parameters should be monitored for at least one year to determine seasonal fluctuations. Analyzed chemical species include Ca, Mg, Na, K, SO_4, HCO_3 (alkalinity), Cl, Fe, Mn, Ba, F, NH_4, NO_2, and NO_3. Redox-sensitive species include Fe, Mn, C, S, and N. Frequently Br is also determined because it can be used as a tracer during performance monitoring.

Data for performance monitoring are concentrations of dissolved species and solid phase data. Principal field parameters are pH, Eh, and DO again. In properly functioning PRB, pH values are high, Eh values are low, and DO is low or below detection limit (O'Hannesin and Gillham, 1998). Decreasing pH (Kamolpornwijit et al., 2003) is together with detectable concentrations of chlorinated solvents at downgradient side of the PRB a sign of decreasing performance. Possibility of by-passing PRB by contaminated ground water can be tested by tracer test. In that case a tracer like bromide or rhodamine WT is injected to wells upgradient of the PRB and recovery of the tracer is monitored in wells downgradient of the PRB.

Solid phase samples are acquired by drilling of inclined wells into PRB. Samples are analyzed by X-ray diffraction, SEM-EDX, and reflected light microscopy. Selective extractions are performed to determine the amount of carbonate minerals (calcite, siderite) and iron monosulfide in solid phase. Build-up of biomass related to the presence of sulfate reducing bacteria (SRB) can be estimated using application of phospholipid fatty acids (PLFA). This is indirect indication based on relation of certain PLFA's to SRB bacterial strains (Wilkin et al., 2003).

Application of ^{13}C for Determination of PRB Performance

Relatively recent application is the use of ^{13}C isotope in chlorinated solvents. Reductive dechlorination of chlorinated solvents produces enrichment of remaining substrate. This is caused by preferential breaking of ^{12}C bonds with resulting isotopic fractionation. The process can be described by Rayleigh distillation. Basic equation for the process is (Clark and Fritz, 1997)

$$R = R_0 . f^{(\alpha - 1)} \tag{13}$$

where R and R_0 are isotopic ratios after and before fractionation and α is fractionation factor. Concentrations of ^{13}C are expressed in delta notation (Clark and Fritz, 1997),

$$\delta^{13}C = \left[\frac{\left(\frac{^{13}C}{^{12}C} \right)_{sample}}{\left(\frac{^{13}C}{^{12}C} \right)_{standard}} \right] x1000 \tag{14}$$

where standard for ^{13}C is PDB (rostrum of fossil *Belemnitella americana* from the Cretaceous Pee Dee Formation in South Carolina).

After application of delta notation and rearrangement we obtain

$$\Delta\delta = \delta - \delta_0 = \varepsilon . \ln f \tag{15}$$

where δ and δ_0 are concentrations of ^{13}C in delta notation in a chlorinated solvent after and before reductive dechlorination, ε is enrichment factor determined experimentally, and f is the ratio between output and input concentrations of a chlorinated solvent. This means that if δ^{13}C values of a chlorinated compound at PRB entry and at PRB outflow, and input concentration C_{input} of the chlorinated compound are known (Fig. 5), it is possible to calculate a theoretical concentration C_{output} of the compound after reductive dechlorination process:

$$C_{output} = C_{input} \cdot \exp\left(\frac{\delta - \delta_0}{\varepsilon}\right) \tag{16}$$

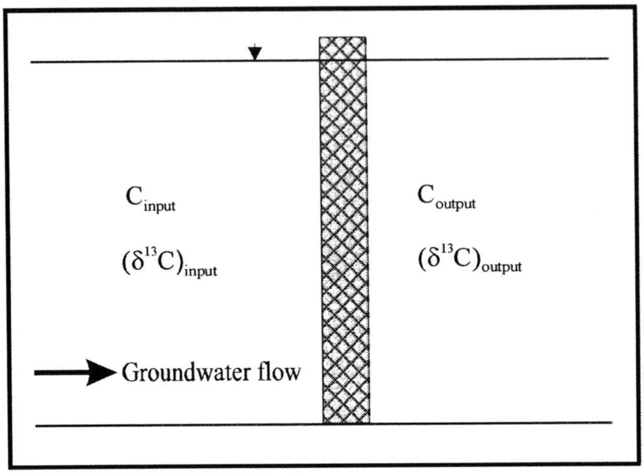

Fig. 5. Calculation of theoretical concentration C_{output} of the compound after reductive dechlorination process according to Eq. (16).

If the theoretical and measured concentrations are the same, then it can be assumed that reduction of concentration of chlorinated compound is due to processes in PRB, and not, for example, caused by plume partially by-passing PRB.

Principal problem is determination of enrichment factor, which is necessary for calculation. Its value is material- and site-specific and has to be determined experimentally (Bill et al., 2001; Shüth et al., 2003).

Geochemical Modeling of PRB

As we state earlier in the text, flow modeling can be used for hydraulic design of PRB. Geochemical and transport modeling represent an extension of flow modeling. Geochemical modeling can be divided broadly into (Zhu and Anderson, 2001; Sracek et al., 2004): (a) speciation modeling, (b) inverse modeling, and (c) forward modeling, and (d) fully coupled geochemical and transport modeling.

Speciation modeling produces only concentrations of dissolved species (free ions and complexes) and saturation indexes with respect to different minerals. When a saturation index (SI) is positive for a mineral, then the mineral may precipitate. Results of speciation modeling can be used for determination of possible mineral phases precipitating in PRB. Then mineralogical investigation can be focused on mineral phases pre-selected on the basis of speciation modeling. Principal problem is common lack of geochemical equilibrium in PRB. This means that many theoretically precipitating minerals will not precipitate due to kinetic constraints.

Inverse modeling is based on the input of initial and final waters and mineral phases. Program generates a set of reactions, which may explain water chemistry changes between hydraulically connected sampling points. Mass transfer coefficients for suggested reactions are also a part of inverse modeling program output.

Forward modeling and forward modeling coupled with transport modeling can reproduce reactions in PRB. In that case initial water is allowed to react with pre-selected mineral phases and final water is compared with field data. Principal problem is again related to the fact that only some reactions in PRB reach equilibrium. This means that kinetic approach has to be applied, but kinetic data are generally limited. Program PHREEQC (Parkhurst and Appelo, 1999), which also has inverse geochemical modeling module, can be used for 1-D simulation of reactive transport in PRB. More complex programs including 3-D simulation of reactive transport were also used, for example by Mayer et al. (2001).

CONCLUSION

Zero-valent iron reacts under anaerobic conditions with halogenated hydrocarbons in water. The halogen derivatives are gradually reduced to non-toxic hydrocarbons. Elemental iron is oxidized to ferrous ion. However, the complete reaction pathway is still uncertain. The reaction is utilizable in chemical decontamination technologies, e.g. in reactive permeable barriers (PRB).

As a simplest approach for the reaction kinetics modeling, first or pseudo first order kinetic equations are frequently used. This description of the kinetics is correct only for a relatively narrow range of concentrations. For a wider range, more sophisticated simulation like power law models, pseudo-stationary models with a controlling surface reaction rate or non-stationary models without the assumption of rate controlling step should be used.

Application of PRB is promising remediation technology. It is a passive remediation method with minimal maintenance.

The principal division of barriers is based on their geometry. Two basic types are continuous barrier and funnel and the gate system. Basic approach for design of PRB follows the steps batch test – column experiment – pilot test. Batch test is used for initial evaluation of material and column test is used for determination of rate constant, which is used for determination of PRB thickness, and also for evaluation of geochemical reactions in reactive material.

Hydraulic aspects of PRB are very important and include determination of the capture zone and residence time in the reactive zone. Ground water flow modeling is frequently performed to accomplish these goals.

The principal problem related to the long term performance of PRB is precipitation of secondary minerals like siderite, calcite, mackinawite, and "green rust" in the reactive material. These processes depend on background ground water chemistry and generally occur in the first decimeters at PRB entry. Precipitated minerals can reduce permeability of PRB, but coating of the surface of reactive material is much more important in the short term.

Performance of PRB requires careful monitoring. The possibility of by-passing PRB by untreated contaminated ground water can be tested by a tracer test and by the application of isotope ^{13}C in chlorinated solvents. In this case, concentrations of ^{13}C at the entry of PRB and at the outflow are compared and a theoretical amount of degraded chlorinated solvents is determined using an enrichment factor specific for a given reactive material.

Geochemical modeling in the speciation mode can be used to select potential mineral phases precipitating in PRB. Inverse geochemical modeling can be used to determine mass transfer in precipitation reactions. The most reliable results are obtained from advanced geochemical modeling combined with transport modeling. In this case, reaction kinetics has to be taken into account.

ACKNOWLEDGEMENT

Financial support of the Grant Agency of the Czech Republic (project 203/03/0925) is greatly appreciated.

REFERENCES

Allen-King R.M., Halket R.M., Burris D.R. (1997): Reductive Transformation and Sorption of cis- and trans-1,2-Dichloroethene in a Metallic Iron–Water System. *Environ. Toxicol. Chem.* 16, 424.

Appelo C.A.J., Postma D. (1996): *Geochemistry, Groundwater and Pollution*, Revised Edition, A.A. Balkema, Rotterdam.

Arnold A.A., Roberts A.L. (2000a): Treatment of Chlorinated Solvents by Nitrogen-Fixing and Nitrate-Supplied Methane Oxidizers in Columns Packed with Unsaturated Porous Media. *Environ. Sci. Technol.* 34, 1794.

Arnold A.A., Roberts A.L. (2000b): Inter- and Intrascpecies Competitive Effects in Reactions of Chlorinated Ethylenes with Zero-Valent Iron in Column Reactors. *Environ. Eng. Sci.* 17, 291.

Baciocchi R., Boni M.R., D'Aprile L. (2003): Characterization of Granular Iron as Reactive Media for TCE Degradation by Permeable Reactive Barriers, *Wat. Air Soil Pollut.* 149, 226.

Beck P., Harries N., Sweeney R. (2001): *Design, Installation, and Performance of a Zero Valent Iron Permeable Reactive Barrier in Monkstown, Northern Ireland*, Claire Technology Demonstration Report TDP3, London.

Benning L.G., Wilkin R.T., Barnes H.L. (2000): Reaction Pathways in the Fe-S System below 100°C, *Chem. Geology* 167, 25.

Bill M., Schüth C., Barth J.A.C., Kalin R.M. (2001): Carbon Isotope Fractionation during Abiotic Reductive Dehalogenation of Trichloroethylene (TCE). *Chemosphere* 44, 1281.

Blowes D.W., Ptacek C.J., Benner S.G., McRae Ch.W.T., Bennet T.A., Puls R.W (2000): Treatment of Inorganic Contaminants Using Permeable Reactive Barriers. *J. Contam. Hydrol.*, 45, 123.

Boronina T., Klabunde K.J., Sergeev G. (1995): Destruction of Organohalides in Water Using Metal Particles: Carbon Tetrachloride/Water Reactions with Magnesium, Tin, and Zinc. *Environ. Sci. Technol.* 29, 1511.

Bourrié G., Trolard F., Génin J.-M.R., Jaffrezic A., Maître V., Abdelmoula M. (1999): Iron Control by Equilibria between Hydroxy-Green Rusts and Solutions in Hydromorphic Soils. *Geochim. Cosmochim. Acta*, 63, 3417.

Burris D.R., Campbell T.J., Manoranjan V.S. (1995): Sorption of Trichloroethylene and Tetrachloroethylene in a Batch Reactive Metallic Iron-Water System. *Environ. Sci. Technol.* 29, 2850.

Campbell TJ., Burris D.R., Roberts A.L., Wells J.R. (1997): Trichloroethylene and Tetrachloroethylene Reduction in a Metallic Iron-Water-Vapor Batch System. *Environ. Toxicol. Chem.* 16, 625.

Cantrell K.J., Kaplan D.I., Wietsma T.W. (1995): Zero-Valent Iron for the In Situ Remediation of Selected Metals in Groundwater. *J. Hazard. Mater.* 42, 201.

Centi G., Parathoner S. (2003): Remediation of Water Contamination Using Catalytic Technologies. *Appl. Catal., B* 41, 15.

Cervini-Silva J., Larson R.A., J. Wu, Stucki J.W. (2002): Dechlorination of Pentachloroethane by Commercial Fe and Ferruginous Smectite. *Chemosphere* 47, 971.

Clark I., Fritz P. (1997): *Environmental Isotopes in Hydrogeology*, Lewis Publishers, Boca Raton, New York.

Dayan H., Abrajano T., Sturchio N.C., Winsor L. (1999): Carbon Isotopic Fractionation during Reductive Dehalogenation of Chlorinated Ethenes By Metallic Iron. *Organic Geochem.* 30, 755.

Deng B., Buris D.R., Campbell TJ. (1999): Reduction of Vinyl Chloride in Metallic Iron-Water Systems. *Environ. Sci. Technol.*33, 2651.

Dombek T., Davis D., Stine J., Klarup D. (2004): Degradation of Terbutylazine (2-chloro-4-ethylamino-6-terbutylamino-1,3,5-triazine), Deisopropyl Atrazine (2-amino-4-chloro-6-ethylamino-1,3,5-triazine), and Chlorinated Dimethoxy Triazine (2-chloro-4,6-dimethoxy-1,3,5-triazine) by Zero Valent Iron and Electrochemical Reduction. *Environ. Pollut.* 129, 267.

Farrell J., Kason M., Melitas N., Li T. (2000): Investigation of the Long-Term Performance of Zero-Valent Iron for Reductive Dechlorination of Trichloroethylene. *Environ Sci. Technol.* 34, 514.

Gavaskar A.R., Gupta N., Sass B.M., Janosy R.J., O'Sullivan D. (1998): *Permeable Barriers for Groundwater Remediation; Design, Construction, and Monitoring.* Battelle Press, New York.

Geiger C.L., Ruiz N.E., Clausen C.A., Reinhart D.R., Quinn J.W. (2002): Ultrasound Pretreatment of Elemental Iron: Kinetic Studies of Dehalogenation Reaction Enhancement and Surface Effects. *Water Res.* 36, 1342.

Génin J.-M.R., Refait P., Bourrié G., Abdelmoula M., Trolard F. (2001): Structure and Stability of the Fe(II)-Fe(III) Green Rust "Fougerite" Mineral and Its Potential for Reducing Pollutants in Soil Solutions, *Appl. Geochem.* 16, 559.

Ghauch A., Gallet C., Charef A., Rima J., Martin-Bouyer (2001): Reductive Degradation of Carbaryl in Water by Zero-Valent Iron. *Chemosphere* 42, 419.

Gibb C., Satapanajaru T., Comfort S.D., Shea P.J. (2004): Remediating Dicamba-Contaminated Water with Zerovalent Iron. *Chemosphere* 54, 841.

Gillham R.W., O'Hannesin S.F. (1994): Enhanced Degradation of Halogenated Aliphatics by Zero-Valence Iron, *Groundwater* 32, 958.

Gotpagar J.K., Grulke E.A., Bhattacharyya D. (1998): Reductive Dehalogenation of Trichloroethylene: Kinetic Models and Experimental Verification. *J. Hazard. Mater.* 62,243.

Gotpagar J., Lyuksyutov S., Cohn R., Grulke E., Bhattacharya D. (1999): Reductive Dehalogenation of Trichloroethylene with Zero-Valent Iron: Surface Profiling Microscopy and Rate Enhancement Studies. *Langmuir* 15, 8412.

Hassan S.M. (2000): Reduction Of Halogenated Hydrocarbons in Aqueous Media: I. Involvement of Sulfur in Iron Catalysis. *Chemosphere* 40, 1357.

Hundal L.S., Singh E.L., Bier E.L., Shea P.J., Comfort S.D., Powers, W.L. (1997): Removal of TNT and RDX from Water and Soil using Iron Metal. *Environ. Pollut.* 97, 55.

Hwang I., Batchelor B. (2002): Reductive Dechlorination of Chlorinated Methanes in Cement Slurries Containing Fe(II). *Chemosphere* 48, 1019.

Cheng I.F., Muftikiian R., Fernando Q., Korte N. (1997): Reduction of Nitrate to Ammonia by Zero-Valent Iron. *Chemosphere* 35, 2689.

Choe S., Lee S.H., Chang Z.Z., Hwang K.Z., Khim J. (2001): Rapid Reductive Destruction of Hazardous Organic Compounds by Nanoscale Fe^0. *Chemosphere* 42, 367.

Janda V., Vasek P., Bizova J., Belohlav Z. (2004): Kinetic Models for Volatile Chlorinated Hydrocarbons Removal by Zero-Valent Iron. *Chemosphere* 54, 917.

Johnson T.L., Fish W., Gorby Y.A., Tratnyek P.G. (1998): Degradation of Carbon Tetrachloride by Iron Metal: Complexation Effects on the Oxide Surface. *J. Contam. Hydrol.* 29, 379.

Johnson T.L., Scherer M.M., Tratnyek P.G. (1996): Kinetics of Halogenated Organic Compound Degradation by Iron Metal. *Environ. Sci. Technol.* 30, 2634.

Kamolpornwijit W., Liang L., West O.R., Moline G.R., Sullivan A.B. (2003): Preferential Flow Path Development and Its Influence on Long-Term PRB Performance: Column Study, *J. Contam. Hydrology* 66, 161.

Lampron K.J., Chiu P.C., Cha D.K. (2001): Reductive Dehalogenation of Chlorinated Ethenes with Elemental Iron: The Role of Microorganisms. *Water Res.* 35, 3077.

Liang L., Korte N., Gu B., Puls R., Reeter C. (2000): Geochemical and Microbial Reactions Affecting the Long-Term Performance of In Situ `Iron Barriers'. *Adv. Environ. Res.* 4, 273.

Lien H.L., Zhang W. (1999): Transformation of Chlorinated Methanes by Nanoscale Iron Particles. *J. Environ. Engr.* 125, 1042.

Lien H.-L., Zhang W. (2002): Enhanced Dehalogenation of Halogenated Methanes by Bimetallic Cu/Al. *Chemosphere*, 49, 371.

Lien H.-L., Zhang W.-X. (2001): Nanoscale Iron Particles for Complete Reduction of Chlorinated Ethenes. *Colloids and Surfaces A: Physicochemical and Engineering Aspects* 191, 97.

Loraine G.A. (2001): Effects of Alcohols, Anionic And Nonionic Surfactants on the Reduction of PCE and TCE by Zero-Valent Iron. *Water Res.* 35, 1453.

Mackay D.M., Cherry J.A. (1989): Ground Water Contamination: Pump-and-Treat Remediation. *Envir. Science. Technol.* 23, 630.

Mackenzie P.D., Horney D.P., Sivavec T.M. (1999): Mineral Precipitation and Porosity Losses in Granular Iron Columns. *J. Hazard. Mater.* 68,1.

Matheson L.J., Tratnyek P.G. (1994): Reductive Dehalogenation of Chlorinated Methanes by Iron Metal, *Environ. Sci. Technol.* 28, 2045.

Mayer K.U., Blowes D.W., Frind E.O. (2001): Reactive Transport Modeling of Groundwater Remediation by an In-Situ Reaction Barrier for the Treatment of Hexavalent Chromium and Trichloroethylene, *Water Resour. Res.* 37, 3091.

McDonald M.G., Harbaugh A.W. (1988): *A Modular Three-Dimensional Finite-Difference Ground-Water Flow Model: Techniques of Water Resources Investigations of the United States*, U.S. Geological Survey, Book 6.

McGuire M.M., Carlson D.L, Vikesland P.J., Kohn T., Grenier A.C., Langley L.A., Roberts A.L., Fairbrother D.H. (2003): Applications of Surface Analysis in the Environmental Sciences: Dehalogenation of Chlorocarbons with Zero-Valent Iron and Iron-Containing Mineral Surfaces. *Anal. Chim. Acta* 496, 301.

Monson S.J., Ma L., Cassada D.A., Spalding R.F. (1998): Confirmation and Method Development for Dechlorinated Atrazine from Reductive Dehalogenation of Atrazine with Fe0. *Anal. Chim. Acta*, 373, 153.

Morales J., Hutcheson R., Cheng I.F. (2002): Dechlorination of Chlorinated Phenols by Catalyzed and Uncatalyzed Fe(0) and Mg(0) Particles. *J. Hazard. Mater*. 90, 97.

Odziemkowski S., Schuhmacher T.T., Gillham R.W., Reardon E.J. (1998): Mechanism of Oxide Film Formation on Iron in Simulating Groundwater Solutions: Raman Spectroscopic Studies. *Corrosion Science* 40, 371.

O'Hannesin S.F., Gillham R.W. (1998): Long-Term Performance of an In-Situ "Iron Wall" for Remediation of VOCs, *Groundwater*, 36, 164.

Orth W.S., Gillham R.W. (1996): Dechlorination of Trichloroethene in Aqueous Solution Using Fe^0. *Environ. Sci. Technol*. 30, 66.

Pankow J.F., Cherry J.A. (1996): *Dense Chlorinated Solvents and other DNAPLs in Groundwater: History, Behavior, and Remediation*, Waterloo Press, Waterloo.

Parkhurst D.L., Appelo A.A.J. (1999): *User's guide to PHREEQC (version 2) - A Computer Program for Speciation, Batch Reaction, One Dimensional Transport, and Inverse Geochemical Modeling*, U.S. Geological Survey Water Resources Investigation Report 99-4259.

Pollock D.W. (1989): *Documentation of Computer Program to Compute and Display Pathlines Using Results from the U.S. Geological Survey Modular Three Dimensional Finite-Difference Modular Three-Dimensional Ground-Water Flow Model*, U.S. Geological Survey Open File Report 89-381.

Reardon E.J., (1995): Anaerobic Corrosion of Granular Iron: Measurement and Interpretation of Hydrogen Evolution Rates, *Environ. Sci. Technol*. 29, 2936-2945.

Ritter K., Odziemkowski M.S., Gillham R.W. (2002): An In Situ Study of the Role of Surface Films on Granular Iron in the Permeable Iron Wall Technology, *J. Contam. Hydrology* 55, 87.

Ritter K., Odziemkowski M.S., Simpgraga R., Gillham R.W., Irish D.E. (2003): An In Situ Study of the Effect Of Nitrate on the Reduction of Trichloroethylene by Granular Iron, *J. Contam. Hydrology* 65, 121.

Roberts A.L., Totten L.A., Arnold W.A., Burris D.R., Campbell T.J. (1996): Reductive Elimination of Chlorinated Ethylenes by Zero-Valent Metals. *Environ. Sci. Technol*. 30, 2654.

Sayles G.D., You G., Wang M., Kupferle M.J. (1997): DDT, DDD, and DDE Dechlorination by Zero-Valent Iron. *Environ. Sci. Technol*. 31, 3448.

Shikaze S. (1996): *3-D Numerical Modeling of Groundwater Flow in the Vicinity of Funnel-and-Gate Systems, ARA-TR-96-5286-1*, Prepared by Applied Research Associates Inc. for U.S. Air Force, Tyndall Air Force Base, FL, USA.

Schäfer D., Köber R., Dahmke A. (2003): Competing TCE and cis-DCE Degradation Kinetics by Zero-valent Iron--Experimental Results and Numerical Simulation. *J. Contam. Hydrol.* 65, 183.

Schreier C.G., Reinhard M. (1994): Transformation of Chlorinated Organic Compounds by Iron and Manganese Powders in Buffered Water and in Landfill Leachate. *Chemosphere* 29, 1743.

Schüth C., Bill M., Barth J.A.C., Slater G.F., Kalin R.M. (2003): Carbon Isotope Fractionation during Reductive Dechlorination of TCE in Batch Experiments with Iron Samples from Reactive Barriers. *J. Cont. Hydrol.* 66, 25.

Slater G.F., Lollar B.S., King R.A., O'Hannesin S. (2002): Isotopic Fractionation during Reductive Dechlorination of Trichloroethene by Zero-Valent Iron: Influence Of Surface Treatment. *Chemosphere* 49, 587.

Sracek O., Bhattacharya P., Jacks G., Gustafsson J.P., Bromsen von, M. (2004): Behavior of Arsenic and Geochemical Modeling of Arsenic Enrichment in Aqueous Environments. *Appl. Geochem*. 19, 169.

Starr R.C., Cherry J.A. (1994): In Situ Remediation of Contaminated Ground Water: The Funnel-and-Gate-System. *Groundwater*, 32, 465.

Su C.W., Puls R. (1999): Kinetics of Trichloroethene Reduction by Zerovalent Iron and Tin: Pretreatment Effect, Apparent Activation Energy, and Intermediate Products. *Environ. Sci. Technol.* 33, 163.

Sweeney K.H., Fischer J.R. (1972): *Reductive Degradation of Halogenated Pesticides*, U.S. Patent No. 3,640,821.

Therrien R., Sudicky E. (1995): Three Dimensional Analysis of Variably-Saturated Flow and Solute Transport in Discretely-Fractured Porous Media. *J. Contam. Hydrology* 23, 1.

U.S. EPA (1998): *Permeable Reactive Barrier Technologies for Contaminant Remediation.* www.epa.gov/ada, Publications, Issue Papers.

Vasek P. (2002): *Removal of Chlorinated Hydrocarbons from Water By Zero-Valent Iron.* Thesis, Institute of Chemical Technology, Prague.

Weerasooriya R., Dharmasena B. (2001): Pyrite-Assisted Degradation of Trichloroethene (TCE). *Chemosphere* 42, 389.

Wilkin R.T., Puls R.W., Sewell G.W. (2003): Long-Term Performance of Permeable Reactive Barriers Using Zero-Valent Iron: Geochemical and Microbiological Effects. *Groundwater*, 41, 493.

Wüst W.F., Köber R., Schlicker O., Dahmke A. (1999): Combined Zero- and First-Order Kinetic Model of the Degradation of TCE and cis-DCE with Commercial Iron. *Environ. Sci. Technol.* 33, 4304.

Zhu C., Anderson G. (2001): *Environmental Applications of Geochemical Modeling*, Cambridge University Press, Cambridge.

In: Water Pollution: New Research
Editor: A. R. Burk, pp. 181-200

ISBN: 1-59454-393-3

Chapter 9

PROTISTS AS TOOLS FOR ENVIRONMENTAL BIOMONITORING: IMPORTANCE OF CHOLINESTERASE ENZYME ACTIVITIES

Maria Umberta Delmonte Corrado*[*]*, Francesca Trielli*[*]*,
***Andrea Amaroli*[*]*, Marzia Ognibene*[*] *and Carla Falugi*[**]**

[*]Department for the Study of the Territory and Its Resources, University of Genoa, Corso Europa 26, I-16132 Genova, Italy

[**]Department of Experimental, Environmental and Applied Biology, University of Genoa, Viale Benedetto XV 5, I-16132 Genova, Italy

ABSTRACT

There is increasing experimental evidence that protists can be exploited as excellent tools for environmental biomonitoring, either as bioindicators of pollution or bioassays to evaluate the effects of toxic compounds. Protists exhibit the fundamental features required to perform rapid biotests with a high degree of reproducibility. Due to their nature as single-cell eukaryotic organisms, they are able to respond directly to environmental stimuli, and multiply through short cell cycles. Extensive efforts are currently devoted to enlarging the panel of organismic systems sensitive to environmental perturbations, such as the occurrence of neurotoxic drugs, which are largely employed on agricultural sites for their insecticide properties. It has long been known that the first target of such toxic compounds is the inhibition of cholinesterase (ChE) enzyme activities, which in turn represents their specific biomarker. In this context, the identification of ChE activities both in the freshwater ciliate, *Paramecium primaurelia,* and the slime mould *Dictyostelium discoideum*, appears interesting, as the sensitivity of these enzyme activities to the presence of neurotoxic drugs allows us to develop a biomonitoring system that exploits these single-cell eukaryotes for the pre-chemical screening of freshwater and moist environments. The ChE activities have been assumed to be involved in cell-cell interactions taking place in *Paramecium* conjugation, as well as in *Dictyostelium* cell migration and aggregation. In fact, the ChE activity inhibition caused by exposure to neurotoxic drugs has been found to perturb mating cell pairing in

P. primaurelia and cell streaming in *D. discoideum*. The alteration to these essential processes for the development in the respective life cycles could result in important ecological effects on natural microbial communities.

INTRODUCTION

The possible damage caused by neurotoxic drugs is now well known. However, only recently the increasing sensitivity to environmental problems led to intensified studies of this kind of pollution present in rural areas, where neurotoxic drugs are employed annually, largely for pest control. These compounds are poisons, able to interfere with the metabolism of their target organisms, thus affecting their central nervous systems, parasympathetic endings, pre-ganglion synapses, and myoneural junctions. Acute intoxication can cause death from respiratory insufficiency. Our extensive knowledge of the effects of neurotoxic drugs is due to their original application as components of nervine gas during both World Wars [1]. The main toxicity of the neurotoxic drugs comes from the inhibition of cholinesterase (ChE) activity, the lytic enzyme of the cholinergic system, as reported in invertebrates such as *Channa punctatus* [2], *Lepomis macrochirus* [3], *Carcinus maenas* [4], *Mytilus galloprovincialis* and *Corbicula fluminea* [5]. Therefore, the inhibition of ChE activity represents a stress index, i.e. a specific biomarker of the presence of neurotoxic drugs [6]. As a consequence, the evaluation of ChE activity is a very powerful tool to be exploited for the assessment of the risk related to the use of neurotoxic drugs.

The most utilised neurotoxic drugs are organophosphate (OP) and carbamate (CA) compounds. The OPs are ester or thiol derivatives of phosphoric acid, phosphonic acid, phosphinic acid, and phosphoramide acid. They bind cell membranes and are soluble in lipids, where they are stored. The chemical structure of OPs is shown in Fig. 1A. Generally, R_1 and R_2 are aryl or alkyl groups, while X includes a wide range of groups, such as the halogen, aliphatic, aromatic, and heterocyclic ones. The OP compounds destroy ChE activity by phosphorylating a specific serine residue within the ChE catalytic centre, forming 0,0-dialkylphosphoserine [7]. It is worth noting that the OPs only phosphorylate the serine of the enzyme active site and, under these conditions, the ChE is unable to hydrolyze the choline esters.

A

B

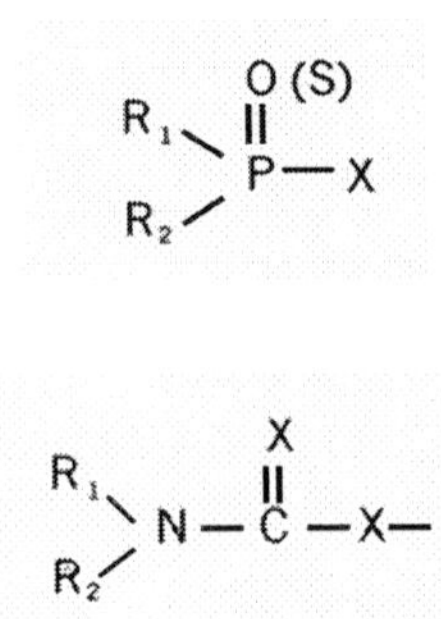

Fig. 1. Chemical structure of organophosphate (**A**) and carbamate (**B**) compounds. (See text for details).

The CAs are derivatives of carbamic acid, are soluble in water and have a short life span. The chemical structure of CAs is represented in Fig. 1B. Generally R_1 and R_2 are organic radicals or H atoms, X can be O_2 or S, and R_3 can correspond to an organic radical or a metal. The CA compounds exert insecticide activity if R_1 is a methyl group and R_2 is an H atom, herbicide activity if R_1 is an aromatic group, fungicide activity if R_1 is a component of benzimidazole [8]. Like OPs, CAs act on the serine within the ChE catalytic center through a carbamylation reaction. However, unlike OPs, the inhibition of ChE activity by CAs is a reversible process [9].

PROTISTS AS TEST ORGANISMS IN ENVIRONMENTAL BIOMONITORING

It is generally agreed that toxicological bioassays offer the possibility of testing the response of organisms to environmental modifications caused by pollutants under controlled conditions. From this point of view, it is of increasing interest to identify a panel of organisms displaying direct and sensitive response mechanisms to environmental perturbations. For this purpose, the sea urchin is a simple model [10, 11, 12], likewise chick and zebrafish embryos, although their organismic complexity makes interpreting the results more difficult. Metazoa cells are also used for bioassays. However, it should be underlined that they are cultured under conditions differing from those occurring in nature and, therefore, their experimental responses will be not directly related to those occurring under natural conditions. On the other hand, it is now widely believed that protists can represent excellent tools to be exploited as test organisms for the pre-chemical screening of environmental sites because they can provide useful information for identifying those situations requiring a close investigation at an early stage [13, 14, 15, 16, 17, 18]. Many studies related to the toxic effects of pollutants on protists have been based on heavy metals, such as Hg, Cd, Zn, Pb, and Cu, because their concentrations have increased to toxic levels in various ecosystems in recent years, due to anthropic activity [19, 20, 21, 22, 23, 24, 25, 26]. Furthermore, several protist species are represented in aqueous and soil microecosystems [27], so they are utilized as biological indicators to assess pollution when their presence or absence can be determined by toxic environmental conditions [15, 28]. Recently, genetically engineered microorganisms have been used in the development of toxicity bioassays. The most widely accepted microbial toxicity bioassay is based on the measurement of a decrease in light emission as function of environmental sample concentration by the wild type luminescent bacterium *Vibrio fischeri*, following a short term exposure to the environmental sample [29]. A drawback often mentioned with regard to this assays is the marine origin of the test organism. Therefore, other organisms have been modified to be used as bioindicators such as the freshwater ciliate *Tetrahymena thermophila*. A promising approach that has been developed on this organism is based on the fusion of a reporter system to selected promoters of different stress response genes, such as metallothionein and Hsp70 [30] genes, for which the expression has been well characterized in *Tetrahymena* under exposure to different environmental contaminants [31, 32, 33].

Protists are particularly useful as test organisms in environmental biomonitoring due to their fundamental features. One basic feature is their nature as an eukaryotic cell/organism, so they exhibit a relatively simple organization and a high degree of specialisation. As organisms, protists respond directly to environmental stimuli, thus behaving like animals as selection units towards environmental challenges. However, as single-cells exposing their receptors directly to the external environment, protists are more sensitive to environmental modifications than the cells of higher organisms that have differentiated and organized complex structures, such as organs and apparatuses, so that cell responses appear to be mediated by different functions. In addition, the absence of a cell wall in the vegetative stage allows protists to respond faster to stimuli than bacteria and yeasts. On the other hand, due to their nature as the primitive eukaryote ancestors of all metazoa, the responses of protists can be related to those of higher organisms. Actually, recent genome studies carried out on *Paramecium* and *Tetrahymena*, commonly employed as test organisms in environmental biomonitoring, highlighted that these ciliates share a high degree of functional conservation with human genes [34, 35]. Furthermore, protists, like microorganisms generally, multiply through short cell cycles, so it is possible to study the effects of toxic compounds on a large and genetically homogeneous cell population in a short time period. A further advantage of protists is the fact that several species can be cultured in the laboratory under seemingly natural conditions, thus their biological responses are more reliable. Finally, protist cultivation is not expensive in terms of human and technical resources.

Cholinergic Molecules

The cholinergic system has been extensively studied and it is known that the cholinergic molecules play the role of specialised transmitters of nervous signals in myoneural junctions, regulating the intercellular messages mediated by variations in the ionic content of the cytoplasm. The cholinergic transmitter system includes the signal molecule acetylcholine (ACh) able to bind its receptors on the surface of target cells. The activity of the ACh lytic enzyme acetylcholinesterase (AChE) modulates the reception function, by hydrolyzing the receptor-bound ACh, thus making the receptor fit for another nervous signal. Therefore, the presence of ACh is always associated with the presence of its lytic enzyme. More generally, the ChEs refer to enzymes able to hydrolyze ACh and other choline esters into their respective components, choline and acetate, or butyrate or propionate [36]. ACh is synthesized by choline acetyltransferase (ChAT), which catalyzes the acetylation of choline by acetyl coenzyme-A (acetyl-CoA) [37], the presence of which was detected in nervous tissues and in numerous non-neuronal tissues as well [38, 39].

Although it is well known that ACh plays a very important role in the transmission of nervous signals, as long ago as 1963, Whittaker [40] stated that ACh is present in non-neuronal tissues and it is so widely expressed in nature as to suggest that it has possible non-nervous functions as well. During the last 60 years, considerable evidence has been accumulated to show that non-neuronal ACh, i.e. independent of neurons [39], is a phylogenetically very old molecule that appeared early in evolutionary history. Actually, ACh has been detected in prokaryotes [41] and in primitive eukaryotes, such as protists [42], sponges [43], and platyhelmints [44], as well as in primitive plants [45]. These results make it possible to denote ACh 'the universal cytotransmitter' [46].

More recent results have shown that non-neuronal ACh can play a multifunctional role [47] and, interacting with other signal molecules, be involved in the modulation of basic cell functions, such as multiplication, differentiation, cell-cell communication, and gamete interaction [39, 48, 49, as reviews]. In the early phases (gastrula) of sea urchin development, ACh could play a role in regulating the embryo fission rate and morphogenetic movements [50, 51]. It has been proposed that during development, the activity of neurotransmitters is involved in communication between all types of cells, so that the presence of ACh and AChE outside the adult nervous system can be explained as residuals of their wide embryo distribution [52, 53].

Two classes of ACh receptors are known to play different roles in excitable cells, even though activated by the same signal molecule. They are the so-called nicotinic receptors (nAChRs) that are nicotine-sensitive, and the so-called muscarinic receptors (mAChRs) that are muscarine-sensitive. Upon ACh excitation, the former allow the rapid transmission of the nervous stimulus by gating Na^+ channels, so causing sudden membrane depolarization [54]; the latter, associated with G-proteins, trigger a transduction cascade that causes the release of secondary messengers [55, 56]. Both the nAChRs and mAChRs are known to be activated or blocked by cholinomimetic drugs. Nicotine and D-tubocurarine are the agonist and antagonist drugs, respectively, of the nAChRs, while carbamylcholine and atropine are the agonist and antagonist drugs, respectively, of the mAChRs.

Depending on the species, different ChEs have been detected and discriminated on the basis of their catalytic properties, according both to the substrate cleaving their activity and to their sensitivity to specific inhibitors [57, 58]. The AChE, also referred to as 'true' AChE (AChE, E.C. 3.1.1.7) present in higher organisms with specialised synapses [59], hydrolyzes ACh with high affinity, and the substrate acetyl-β-methyl thiocholine iodide (AcTChI) as well. AChE enzyme activity is inhibited by eserine, a carbamate compound, and, more specifically, by BW 284c51, a phenol ester. Other choline esters, such as butyrylcholinesterase (BChE, E.C. 3.1.1.8) and propionylcholinesterase (PrChE), are also referred to as 'pseudocholinesterases'. The best-known molecule of aspecific ChE activity is BChE that preferentially hydrolyzes butyrylcholine (BCh) and the substrate butyryl thiocholine iodide (BTChI) [58]. It has been suggested that BChE arose from mutation of the gene encoding AChE late in evolutionary history, so that the duality of AChE and BChE is likely to be specific to vertebrates, probably originating in the ancestors of gnatostomes [59]. There is little information on PrChE in the literature except for its molecular forms that have been characterized in primitive organisms [58, 60]. This enzyme, detected in bovine brain caudate nucleus [61], is practically absent in higher organisms, and this could be the reason that PrChE has not been extensively studied and its functions have not been investigated so far. According to Talesa et al. [58], PrChE is likely to be an ancestral form, able to cleave a wider range of choline esters other than AChE, and propionyl tiocholine iodide (PrTChI) as a preferential substrate. Both BChE and PrChE enzyme activities are inhibited by eserine and, more specifically, by iso-OMPA, a phosphoramide compound.

CHOLINERGIC-LIKE MOLECULES IN PROTISTS

There is little information on the presence and the functional role of molecules related to the cholinergic system in protists. It has been suggested that cholinergic-like molecules can be involved in the mechanism of prey adherence in the ciliated suctoria [62], and in regulating both ciliary motility and cell division in the ciliate *Paramecium* [63]. Moreover, the presence of different esterases during the life cycle of the ciliate *Euplotes crassus* has been cytochemically demonstrated [64]. Recently, we have shown the presence of molecules belonging to the cholinergic signaling system in *P. primaurelia* [65, 66, 67] and in the sarcodine *Dictyostelium discoideum* [68, 69]. Molecules related to the GABAergic system have also been identified in *P. primaurelia* [70]. The detection of neurotransmitter molecules in primitive non-nervous eukaryotes supports the hypothesis of a progression in their role, from being both intracellular and intercellular messengers of the relatively slow cell-cell communications of protists, to being intercellular messengers for the rapid cell-cell communications between the highly specialised synaptic structures of metazoa [71].

In *D. discoideum* the presence of a serine esterase has been found to show a strong sequence identity to *Torpedo* AChE [72]. *D. discoideum*, also referred to as slime mould, is a paradigmatic organism from an evolutionary point of view as it exhibits morphological characters typical of a very primitive eukaryote and a particularly complex developmental cycle including the phases of single cell multiplication and aggregated cell differentiation. During the former phase, single-cell amoebae feeding on bacteria grow and multiply by binary fission. The latter phase starts from depletion of the food supply that triggers chemotaxis of amoebae. The amoebae communicate with each other when stimulated by the release of cyclic AMP that acts as an intercellular signal molecule and aggregate in streams, giving rise to the first multicellular stage in which cell differentiation takes place. Such a process culminates in the formation of the fruiting body, anchored to the substrate by the basal stalk, containing two differentiated cell types: the 'somatic' stalk cells unable to multiply, and the apical spores, which are 'generative' cells able to reproduce under suitable environmental conditions (Fig. 2) [73, 74, as reviews].

We have histochemically demonstrated that there are three ChE activities in *D. discoideum* single-cell amoebae that are able to cleave the substrates AcTChI, PrTChI, and BTChI, respectively [68], giving rise to brown precipitates [75]. For the spectrophotometric evaluation, ChE activities were recorded at λ = 412 nm, using a Uvikon 930 spectrophotometer (Kontron Instruments), during the first 10 min after exposure to the specific substrate according to the method proposed by Ellman et al. [76] with modifications. Measurements were carried out at 25 °C, pH 8.0. The enzyme activity was expressed as ChE units (U)/min/µg total protein (TP) content, and the TP was evaluated spectrophotometrically according to the Millon-Nasse method [77]. The main ChE activity was able to hydrolyze the PrTChI substrate, whereas the ChE activity cleaving the AcTChI substrate was significantly lower, and that cleaving the BTChI substrate was very low. AChE-, PrChE-, and BChE-like molecules of the membrane fraction from single-cell amoebae were also checked by separating them on an 8% polyacrilamide running gel, according to the non-denaturing electrophoresis method proposed by Ornstein and Davis [78]. The AcTChI, PrTChI, and BTChI were used as substrates and *Electrophorus electricus* AChE was employed as a control.

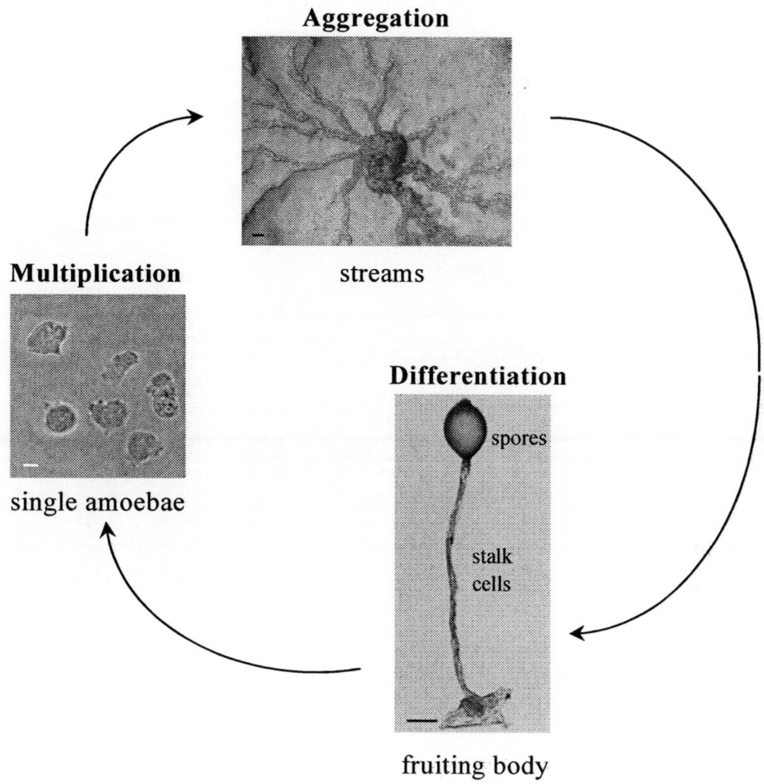

Fig. 2. A simplified scheme of the developmental cycle of *Dictyostelium discoideum*. The life cycle starts from the multiplication phase, during which single-cell amoebae feed on bacteria, grow, and reproduce by binary fission. The next phase starts from the depletion of food supply that triggers aggregation of the amoebae in streams to form a multicellular aggregate in which cell differentiation takes place. The final phase culminates in the development of the fruiting body, containing the apical spores able to reproduce under suitable environmental conditions, and the stalk cells unable to multiply. White bar = 10 μm; black bars = 300 μm.

The electrophoretic pattern showed that the PrTChI substrate was hydrolyzed by a single enzyme activity and the AcTChI substrate by two enzyme activities, one migrating like the 260 kDa molecular form of *E. electricus* AChE [79], the other corresponding to the molecular form revealed by the PrTChI substrate. This finding is consistent with the above-mentioned hypothesis advanced by Talesa et al. [58] that PrChE is an ancestral form able to cleave a wider range of choline esters than AChE. No BChE activity was detected using BTChI as a substrate. To characterize the PrChE, the highest ChE activity found in *D. discoideum* single-cell amoebae, the effects of specific inhibitors of ChE activities were evaluated spectrophotometrically. Exposure to 10^{-5} M iso-OMPA, a specific inhibitor of pseudocholinesterases, completely blocked the PrChE activity, while exposure to 10^{-5} M BW284c51, a specific inhibitor for true vertebrate cholinesterase, caused a significant decrease in PrChE activity. The sensitivity of PrChE activity to variations in pH and temperature was also assessed. As the pH increased from 5.0 to 10.0 at 25 °C, the enzyme activity showed gradually increasing values, the lowest one corresponding to pH 5.0 and the highest one to pH 10.0 (Fig. 3A). In a temperature interval ranging from 5 °C to 37 °C, the PrChE activity evaluated at pH 8.0 showed the lowest values between 5 °C and 20 °C, followed by significantly increased values at 25 °C, 30 °C and 37 °C (Fig. 3B).

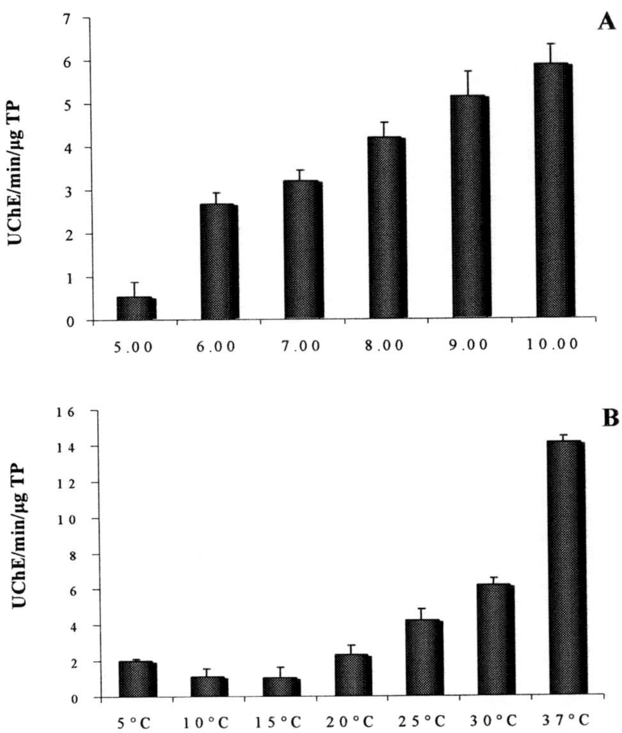

Fig. 3. Spectrophotometric evaluation of the cholinesterase (ChE) activity of *D. discoideum* single-cell amoebae, able to cleave the substrate propionyl thiocholine iodide, surveyed at increasing pH values at 25 °C (**A**) and at increasing temperature values at pH 8.0 (**B**). The enzyme activity was expressed as units (U) of ChE/min/μg total protein (TP) content. Means of three experiments and standard deviation. Student's *t*-test was used to compare the means.

The presence of molecules belonging to the cholinergic signaling system was also detected in the developmental cycle of *P. primaurelia*, using different methodological approaches. As receptor-bound ACh is rapidly hydrolyzed by AChE [80] its detection is very difficult to demonstrate; therefore, we showed immunocytochemically the presence and histochemically the activity of the biosynthetic ACh enzyme ChAT (E.C. 2.3.1.6). The presence of ChAT activity was found to be related to the phases of the developmental cycle. In *P. primaurelia*, the life cycle (Fig. 4) includes the sequential phases of immaturity, which follows conjugation and is characterized by the cells' inability to undergo sexual processes, and maturity, during which the cells become mating-competent and able to conjugate.

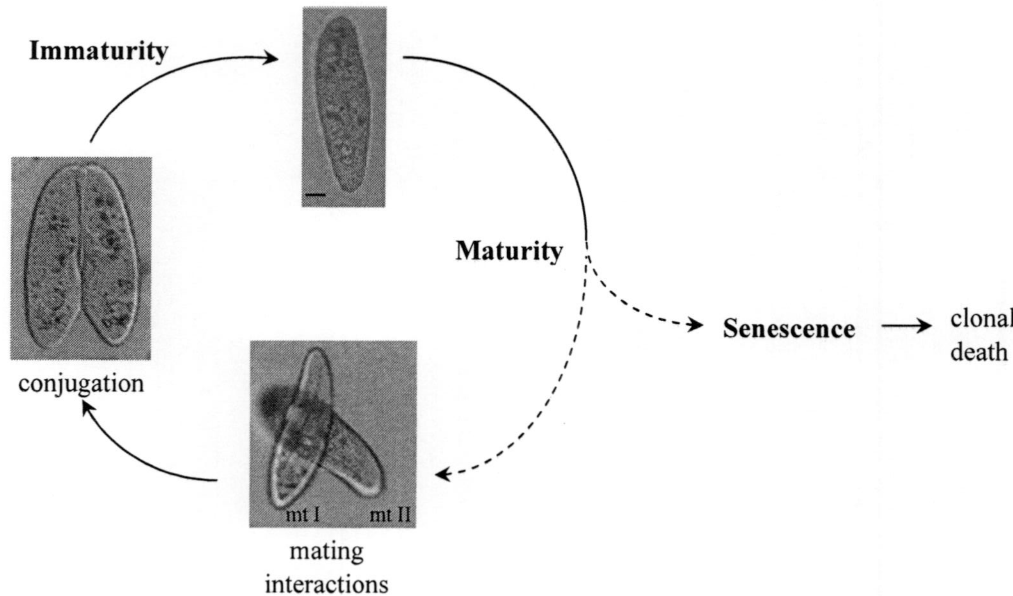

Fig. 4. A simplified scheme of the developmental cycle of *Paramecium primaurelia*. The life cycle includes the phase of immaturity that follows conjugation, and maturity during which the differentiation of mating type (mt) I and mt II cells can take place. After mating interactions, these cells are able to join in conjugation. If conjugation is prevented, clonal senescence occurs and eventually leads to clonal death. Bar = 20 μm.

Conjugation is triggered by cell-cell interactions between differentiated classes of mating-competent cells, usually denoted as mating type I and mating type II, distinguished from one another by means of molecules responsible for the mating-type specificity, and classically referred to as 'mating substances'. Mating interactions take place immediately after the mixing of mating-competent cells of opposite mating types, which clump in clusters and then separate in mating-cell pairs [81, 82, as reviews]. If mature cells are prevented from conjugating, they enter senescence that eventually leads to clonal death. It is worth noting the peculiar effect of conjugation on its clonal progeny. There is sustained evidence that the sexual processes of ciliates not only allow the gene recombination to occur, so increasing the genetic variability of the clonal population, but are also able to reset the clonal age to zero, thus 'rejuvenating' their clonal progeny and avoiding senescence [83]. We detected the presence of ChAT activity in mating-competent cells, mainly localized on the membrane surface; conversely, ChAT activity was not found in immature cells [67]. Therefore, the synthesis of ACh seems to occur at the cortical level in mating-competent cells, thus suggesting a function for the cholinergic system in cell-cell signaling during mating interactions. Immunoblot analysis revealed the presence of immunoreactive bands, ranging in molecular mass from 42 to 133 kDa, as reported for ChAT isolated from higher organisms [84, 85, 86], and distributed differently according to the phase of the developmental cycle.

Two ChE activities able to cleave the substrates AcTChI and PrTChI, respectively, were detected histochemically and spectrophotometrically (Fig. 5). However, the presence and the activity of BChE-like molecules was not revealed by Western blot and non-denaturant electrophoresis, respectively [66], thus supporting the above mentioned hypothesis related to the late occurrence of BChE in evolution [59]. The most represented ChE activity detected in *P. primaurelia* was AChE, typical of higher organisms. On the contrary, *D. discoideum,* which is considered a less-evolved organism than *Paramecium*, showed the ancestral form PrChE as the prevalent ChE in the single-cell amoebae phase [69]. To characterize the pre-eminent ChE found in *P. primaurelia*, AChE, we spectrophotometrically evaluated its sensitivity to exposure to 10^{-5} M iso-OMPA or eserine or BW 284c51, classical inhibitors of ChE activities. A significant decrease in ChE activity was found after incubation with eserine, an inhibitor of both AChE and BChE, and BW 284c51, a specific inhibitor of AChE in different organisms (Fig. 6).

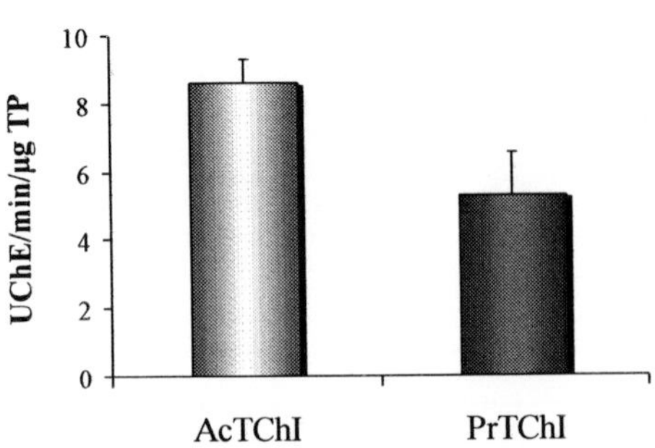

Fig. 5. Spectrophotometric evaluation of cholinesterase (ChE) activities of *P. primaurelia* able to cleave the substrates acetyl-β-methyl thiocholine iodide (AcTChI) or propionyl thiocholine iodide (PrTChI). The enzyme activities were expressed as units (U) of ChE/min/µg total protein (TP) content. Means of three experiments and standard deviation. Student's *t*-test was used to compare the means.

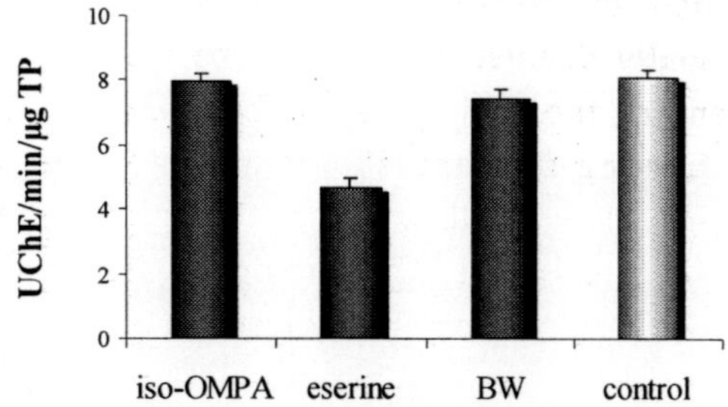

Fig. 6. Spectrophotometric evaluation of cholinesterase (ChE) activity of *P. primaurelia* able to cleave the substrate acetyl-β-methyl thiocholine iodide, in cells exposed to ChE inhibitors (10^{-5} M) and in control cells. The compounds used were iso-OMPA, eserine, and BW284c51. The enzyme activity was expressed as units (U) of ChE/min/μg total protein (TP) content. Means of three experiments and standard deviation. Student's *t*-test was used to compare the means.

The non-denaturant electrophoresis pattern of AChE activity showed that this enzyme activity was due to a 260-kDa molecular form, similar to the membrane-bound tetrameric form present in human erythrocytes [66]. Furthermore, the AChE activity appeared to be localized in different sites, according to the phases of the developmental cycle, being present in the membrane fraction of mating-competent cells and in the cytoplasmic fraction of immature cells. Moreover, in mating-competent cells, molecules immunologically related to mAChRs were detected by immunofluorescence, and nAChR-like molecules were revealed as α-bungarotoxin-binding sites [65]. The presence of AChE activity and both ACh receptors on the surface membrane of mating-competent cells suggested a possible role for AChE in regulating cell-cell interactions mediated by transmitter molecules, such as ACh. Therefore, to investigate whether cholinergic-like molecules were involved in modulating mating interactions, experiments were performed in which mating-competent cells of opposite mating types were exposed to 10^{-4} M carbamylcholine or 10^{-7} M nicotine, which are agonist drugs of the mAChR or nAChR function, respectively, before cell mixing. Similarly, mating-competent cells were exposed to 10^{-4} M atropine or 10^{-7} M D-tubocurarine chloride, which are antagonist drugs of the mAChR and nAChR function, respectively. All incubations with these cholinomimetic compounds appeared to affect mating interactions as mating-cell pairing was inhibited significantly. Afterwards, the analysis was extended to the effects of AChE- or ChAT-inhibitors on mating-cell pairing [67]. Mating-competent cells of opposite mating types were exposed to 10^{-4} M BW 284c51 or to 10^{-4} M eserine, both inhibitors of AChE activity, and to 10^{-3} M Congo Red, inhibiting ChAT activity [38]. It is known that blocking AChE activity prevents AChE from cleaving ACh and leads to the overproduction of ACh at its receptor sites [59]. Both AChE-inhibitors significantly perturbed mating-cell pairing. A possible explanation is that an excess of ACh at its receptor sites can determine either an overexcitatory or an inhibitory effect on ACh receptor function, both actions altering cell functions, as reported in sea urchin spermatozoa [87]. Conversely, blocking ChAT activity and, consequently, ACh synthesis with Congo Red, which is a potent competitive inhibitor of Acetyl-CoA [38], did not significantly affect mating-cell pairing, thus suggesting that ACh has a negative modulatory effect on mating interactions.

Cholinesterase Activity of Protists as a Biomarker of Environmental Stressors

When using ChEs as biomarkers it should be remembered that non-contaminant factors can affect their activity. In the case of employing macroinvertebrates and vertebrates as test organisms for the occurrence of neurotoxic drugs, variations in the ChE activity can be caused by numerous non-contaminant factors, such as type of tissue, species, genetic variation, age, sex, circadian and seasonal rhythms, reproductive state, and endocrine regulation [88]. These interference factors are avoided in the case of employing protists as test organisms because they are single-cell organisms whose asexual reproduction gives rise to phenotypically and genotypically homogeneous cell populations. The possible stress factors interfering with ChE activity are related to the physiological and feeding conditions of the protist cell culture, and to the temperature level. However, these factors can be discounted because of the standard conditions used for growing and culturing protists in the laboratory.

To evaluate the possibility of exploiting the ChE activities detected in *D. discoideum* and in *P. primaurelia* as specific biomarkers for the presence of neurotoxic drugs, we spectrophotometrically tested the sensitivity of these enzyme activities to exposure to OP and CA compounds prior to incubation with their specific substrates. For *D. discoideum* PrChE, the OPs employed were basudin, a formulate insecticide utilized world-wide since 1952, diazinon, a thionophosphate active principle contained (20%) in basudin, and cidial, a dithiophosphate active principle. The CAs were carbaryl and pirimicarb, both active principles. Exposure to the above mentioned neurotoxic drugs, used at a concentration of $1/10^6$ v/v, caused a significant decrease (more than 50%) in PrChE activity when compared to that of the control (Fig. 7). Moreover, a dose-dependent inhibition of PrChE activity was found for basudin concentrations ranging from $1/10^1$ v/v to $1/10^7$ v/v [68]. The enzyme activity tended to zero value for basudin concentrations between $1/10^3$ v/v and $1/10^1$ v/v.

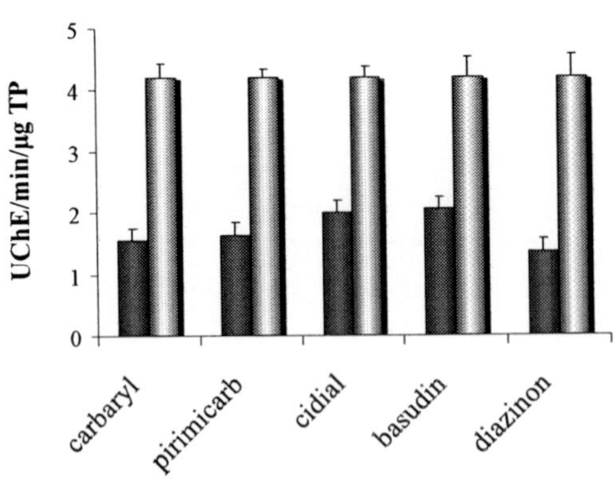

Fig. 7. Spectrophotometric evaluation of cholinesterase activity (ChE) of *D. discoideum* single-cell amoebae able to cleave the substrate propionyl thiocholine iodide, in cells exposed to neurotoxic drugs ($1/10^6$ v/v) (dark bars) and in control cells (white bars). Carbaryl and pirimicarb were employed as carbamates; cidial, basudin, and diazinon were employed as organophosphates. The enzyme activity was expressed as units (U) of ChE/min/µg total protein (TP) content. Means of three experiments and standard deviation. Student's *t*-test was used to compare the means.

The effects of basudin on cell morphology, cell density and cell differentiation were also assessed. A three-day exposure to basudin at a concentration of $1/10^4$ v/v affected the cell morphology as the amoeboid shape was lost and the cell shape tended to take on a circular form. Furthermore, the cell density did not increase, thus suggesting a possible inhibitory effect on cell multiplication. After the three-day exposure to basudin, the single-cell amoebae were transferred to an agar-N plate containing an *Escherichia coli* monolayer, to test the cells' ability to aggregate and differentiate. Unlike the control, which formed a uniform layer of fruiting bodies with well-developed sorocarps after depleting bacteria, the basudin-exposed amoebae only depleted bacteria surrounding the seeding points and were generally unable to aggregate, so that only rare fruiting bodies, with very small sorocarps, were formed.

The sensitivity of both the AChE and PrChE activities detected in *P. primaurelia* to neurotoxic drugs was tested by incubating cells with $1/10^4$ v/v diazinon or carbaryl, an OP or a CA compound, respectively. Both drugs significantly inhibited AChE activity (Fig. 8A), while they did not affect PrChE activity (Fig. 8B). Therefore, unlike PrChE, AChE was considered as a potential biomarker for the presence of neurotoxic drugs. To evaluate the effect of an OP compound on mating interactions, mating-competent cells of opposite mating types were exposed to diazinon at a concentration of $1/10^4$ v/v prior to mixing. Mating-cell pairing was found to be strongly inhibited, as the diazinon-exposed cell populations were unable to pair (unpublished data not shown).

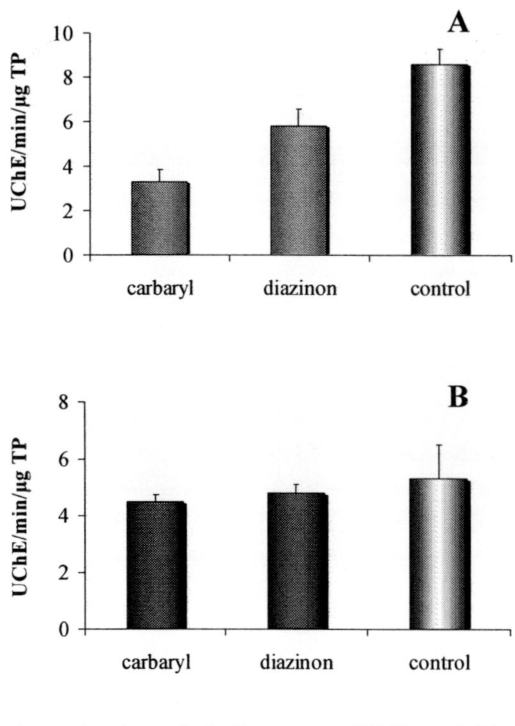

Fig. 8. Spectrophotometric evaluation of cholinesterase (ChE) activities of *P. primaurelia* able to cleave the substrates acetyl-β-methyl thiocholine iodide (**A**) or propionyl thiocholine iodide (**B**), respectively, in cells exposed to neurotoxic drugs ($1/10^4$ v/v), and in control cells. Carbaryl and diazinon were employed as a carbamate and an organophosphate compound, respectively. The enzyme activities were expressed as units (U) of ChE/min/µg total protein (TP) content. Means of three experiments and standard deviation. Student's *t*-test was used to compare the means.

On the whole, the results described in this chapter show that *D. discoideum* PrChE and *P. primaurelia* AChE activities are sensitive to exposure to both OP and CA compounds, even at the low concentrations that are often pollution conditions themselves. This feature should allow us to develop a biomonitoring system that exploits *D. discoideum* and *P. primaurelia* as potential test organisms for the pre-chemical screening of wetland and fresh water environments. In fact, our spectrophotometric method to evaluate the presence of neurotoxic drugs in fresh water sites by measuring ChE activities will shortly be published [89].

A remarkable finding highlighted by our investigations is that the ChE activities detected in *D. discoideum* and *P. primaurelia* seem to be involved in regulating cell-cell signaling, as their inhibition by neurotoxic drugs affects cell-cell interactions leading to cell aggregation in *D. discoideum* and to mating-cell pairing in *P. primaurelia*. The ecological consequences of these effects will be discussed in the following section.

We have recently considered the possibility of utilizing *D. discoideum* as a test organism and its PrChE activity as a biomarker for studying the effects of exposure to time-variant, extremely low-frequency electromagnetic fields (ELF-EMF) (180 μTesla, and 50 Hz). Likewise the ChE activity in the frontal cortex and hippocampus of the rat [90], the PrChE activity in single-cell amoebae was found to be sensitive to this EMF. However, EMF effect was transient, as PrChE activity returned to the control value after transferring the cells to standard conditions. Moreover, a lag in cell aggregation was observed, but it was subsequently recovered, so that the fruiting bodies derived from ELF-EMF-exposed single-cell amoebae were normal both in number and size at the end of the differentiation process (data not shown) [91].

Concluding Remarks

On the whole, the results presented in this chapter allow some considerations. First of all, the above-described effects of basudin and diazinon, inhibiting cell aggregation in *D. discoideum* and mating-cell pairing in *P. primaurelia*, respectively, could have serious ecological consequences. Exposure to neurotoxic drugs appears to block the developmental cycle of *Dictyostelium* as the differentiation of the spore, i.e. the cells maintaining the capacity to reproduce, is prevented. Likewise, neurotoxic drug-exposed mating-competent cells of *Paramecium* are found to be unable to join in pairs and ultimately to conjugate. The inhibition of conjugation, which reset the clonal age to zero 'rejuvenating' its clonal progeny, causes clonal senescence and death. Therefore, neurotoxic drugs could affect the natural balance of whole microbial communities, as the trophic chain could be strongly influenced. As previously reported, protists are a very important component of the microecosystems inhabiting moist and freshwater environments. Numerous protists, such as *Dictyostelium* and *Paramecium*, are bacterivorous and, in turn, represent a food source for higher organisms. A decrease in the protist component of the microbial community would result in an increase in the bacterial component and this event would alter the environment and its subsequent evolution. Therefore, it appears that the clonal health of both *Dictyostelium* and *Paramecium* cell populations may be a valuable 'sentinel' for assessing the environmental balance between the different components of the trophic chain, and the possible risk for human health, as neurotoxic drugs can jeopardize the health of farm-workers and anyone else using the contaminated environment.

Another remark worth noting is the evaluation of the biological effects of the neurotoxic drugs employed in our studies, which makes it possible to detect biological targets of cell metabolism and functions common to primitive and evolved eukaryotes. In particular, we may identify the effects on cell-cell communication mechanisms, which are the basis of the functionality of multicellular organisms, as excellent biomarkers of neurotoxic drugs. These mechanisms are really relevant for animals, included man, as ACh, the signal molecule of the cholinergic system, is a simple, phylogenetically old molecule [39], whose structural features have been conserved throughout evolution. Although other cholinergic molecules, such as ChEs, ChAT, and both mACh and nACh receptors, are proteins and can potentially exhibit structural differences in evolutionarily different organisms, their function is determined and modulated by ACh functions according to a 'standard' mechanism that evokes the same ionic and transduction intracellular responses in both single-cell and multicellular eukaryotes. Moreover, the presence of multiple signal molecule systems found in protists [48, 65, 66, 67, 70] may explain the wide range of effects that neurotoxic drugs (which are often suspected of causing severe neurological problems in animals, including man) have on the transmitter systems through the interactions among the different molecular sets at every level of organismic complexity [92].

In conclusion, it is desirable that the single-cell eukaryote systems utilized in our studies are considered by the international agencies for environment welfare as valuable tools to be exploited for environmental biomonitoring also in view of their sensitivity to environment perturbations affecting cell-cell communications.

REFERENCES

[1] Karczmar A.G., Usdin E., Wills J.H. (1970). Anticholinesterase agents. In: *International Encyclopedia of Pharmacology and Therapeutics*. Pergamon Press, Oxford, pp 249-262.

[2] Ghosh P., Bhattacharya S. (1992). In vivo and in vitro AChE inhibition by metacid-50 and carbaryl in *Channa punctatus* under natural field conditions. *Biomed. Environ. Sci.*, 5, 18-24.

[3] Dutta H., Marcelino J., Richmonds C. (1997). Brain AChE activity and optomotor behavior in bluegills *Lepomis macrochirus* exposed to different concentrations of diazinon. *Arch. Intern. Physiol. Biochim. Biophys.*, 100, 331-334.

[4] Lundebyu A.K., Curtis T.M., Braven J., Depledge M.H. (1997). Effects of the O.P. pesticide, dimethoate on cardiac and AChE activity in the shore crab *Carcinus maenas*. *Aquatic Toxicol.*, 40, 23-36.

[5] Mora P., Fournier D., Narbonne J.F. (1999). Cholinesterase from the marine mussels *Mytilus galloprovincialis* LmK and *M. edulis* L. and from the freshwater bivalve *Corbicula fluminea* Muller. *Comp. Biochem. Physiol. Pharmacol. Toxicol. Endocrinol.*, 122, 353-361.

[6] Sultatos L.G. (1994). Mammalian toxicology of organophosphorous pesticides. *J. Toxicol. Environ. Health*, 43, 271-289.

[7] Kennedy S.W. (1991). The mechanism of organophosphate inhibition of cholinesterase. Proposal for a new approach to measuring inhibition. In: *Cholinesterase-inhibition Insecticides: their Impact on Wildlife and the Environment. Chemicals in Agricolture.* Vol. 2. P. Mineau (ed.). Elsevier Science Publishers, Amsterdam, pp 74-87.

[8] Muccinelli M. (1997). *Prontuario dei Fitofarmaci.* 8th ed. Edagricole-Edizioni Agricole Calderini, Bologna.

[9] Kuhr R.J., Dorough H.W. (1976). Mode of action. In: *Carbamate Insecticides: Chemistry, Biochemistry and Toxicology.* R.J. Kuhr, H.W. Dorough (eds). CRC Press, Cleveland, pp 41-70.

[10] Pagano G., Cipollaro M., Corsale G., Esposito A., Ragucci E., Giordano G.G., Trieff N.M. (1986). The sea urchin: bioassay for the assessment of damage from environmental contaminants. In: *Community Toxicity Testing.* J. Jr Cairns (ed.). American Society for Testing and Materials, Philadelphia, pp 67-92.

[11] Morale A., Coniglio L., Angelini C., Cimoli G., Bolla A., Alleteo D., Russo P., Falugi C. (1998). Biological effects of a neurotoxic pesticide at low concentrations on sea urchin early development. A teratogenic assay. *Chemosphere,* 37, 3001-3010.

[12] Arizzi Novelli A., Volpi Ghirardini A., Giuliani S., Falugi C., Pagano G., Iaccarino M. (2001). Evaluation of coastal water and sediment toxicity towards gametes and embryos from sea urchins using different test procedures. *Biol. Mar. Medit.*, 8, 41-59.

[13] Apostol S. (1973). A bioassay of toxicity using protozoa in the study of aquatic environment pollution and its prevention. *Environ. Res.*, 6, 365-372.

[14] Persoone G., Dive D. (1978). Toxicity tests on ciliates - a short review. *Ecotoxicol. Environ. Saf.*, 2, 105-114.

[15] Lynn D.H., Gilron G.L. (1992). A brief review of approaches using ciliated protists to asses aquatic ecosystem health. *J. Aquatic Ecosys. Health,* 1, 263-270.

[16] Ricci N. (1995). Protozoa: bioindicators of environmental monitoring and pollution control. In: *Pollution and Biomonitoring.* B.C. Rana (ed.). Tata McGraw-Hill Publishing Company, New Delhi, pp 240-269.

[17] Sauvant M.P., Pepin D., Piccinni E. (1999). *Tetrahymena pyriformis*: a tool for toxicological studies. A review. *Chemosphere,* 38, 1631-1669.

[18] Miyoshi N., Kawano T., Tanaka M., Kadono T., Kosaka T., Kunimoto M., Takahashi T., Hosoya H. (2003). Use of *Paramecium* species in bioassays for environmental risk management: determination of IC_{50} values for water pollutants. *J. Health Sci.*, 49, 429-435.

[19] Dini F. (1981). Relationship between breeding systems and resistance to mercury in *Euplotes crassus* (Ciliophora: Hypotrichida). *Mar. Ecol. Prog. Ser.*, 4, 195-202.

[20] Piccinni E., Irato P., Coppellotti O., Guidolin L. (1987). Biochemical and ultrastructural data on *Tetrahymena pyriformis* treated with copper and cadmium. *J. Cell Sci.*, 88, 283-293.

[21] Nilsson J. R. (1989). *Tetrahymena* in cytotoxicology: with special reference to effects of heavy metals and selected drugs. *Eur. J. Protistol.*, 25, 2-25.

[22] Madoni P., Davoli D., Gorbi G. (1994). Acute toxicity of lead, chromium, and other heavy metals to ciliates from activated sludge plants. *Bull. Environ. Contam. Toxicol.*, 53, 420-425.

[23] Viarengo A., Accomando R., Ferrando I., Beltrame F., Fato M., Marcenaro G. (1996). Heavy metal effects on cytosolic free Ca^{2+} level in the marine protozoan *Euplotes*

crassus evaluated by confocal laser scanning microscopy. *Comp. Biochem. Physiol.*, 113C, 161-168.

[24] Ricci N., Luverà G., Cacciatori M., Banchetti R., Lueken W. (1997). The effects of 2 μM Hg^{++} on the ethogram of *Euplotes vannus* (Ciliata, Hypotrichida). *Europ. J. Protistol.*, 33, 63-71.

[25] Coppellotti O. (1998). Sensitivity to copper in a ciliate as a possible component of biological monitoring in the lagoon of Venice. *Arch. Environ. Contam. Toxicol.*, 35, 417-425.

[26] Gutiérrez J.C., Martín-González A., Díaz S., Ortega R. (2003). Ciliates as potential source of cellular and molecular biomarkers/biosensors for heavy metal pollution. *Europ. J. Protistol.*, 39, 461-467.

[27] Fenchel T. (1987). *Ecology of Protozoa: The Biology of Free-living Phagotrophic Protists*. Madison/Springer-Verlag, Berlin.

[28] Nicolau A., Dias N., Mota M., Lima N. (2001). Trends in the use of protozoa in the assessment of wastewater treatment. *Res. Microbiol.*, 152, 621-630.

[29] Quershi A.A., Bulich A.A., Isemberg D.L. (1998). Microtox toxicity test systems - where they stand today. In: *Microscale Testing in Acquatic Toxicology: Advances, Technique and Practice*. P.G. Wells, K. Lee, C. Blaise (eds). CRC Press, Boca Raton, pp 185-199.

[30] La Terza A., Papa G., Miceli C., Luporini P. (2001). Divergence between two Antarctic species of the ciliate *Euplotes*, *E. focardii* and *E. nobilii*, in the expression of heat-shock protein 70 genes. *Mol. Ecol.*, 10, 1061-1067.

[31] Piccinni E., Bertaggia D., Santovito G., Miceli C., Kraev A. (1999). Cadmium metallothionein gene of *Tetrahymena pyriformis*. *Gene*, 234, 51-59.

[32] La Terza A., Barchetta S., Miceli C. (2003). Development of cell lines of ciliated protozoa genetically modified to be used as biosensors of organotin and other contaminants in the water. SIBMA 2003, *Marine and Environmental Biochemistry*, San Benedetto del Tronto.

[33] Dondero F., Cavaletto M., Ghezzi A.R., La Terza A., Viarengo A. (2004). Biochemical characterization and quantitative gene expression analysis of the multi-stress inducible metallothionein from *Tetrahymena thermophila*. *Protist*, 155, in press.

[34] Dessen P., Zaguilki M., Gromadka R., Plattner H., Kissmehl R., Meyer E., Bétermier M., Schultz J.E., Linder J.U., Pearlman R.E., Kung C., Forney J., Satir B.H., Van Houten J.L., Keller A-M., Froissard M., Sperling L., Cohen J. (2001). *Paramecium* genome survey: a pilot project. *Trends Genet.*, 17, 306-308.

[35] Turkewitz A., Orías E., Kapler G. (2002). Functional genomics: the coming of age for *Tetrahymena thermophila*. *Trends Genet.*, 18, 35-40.

[36] Stedman E., Easson L.H. (1932). Cholinesterase: an enzyme present in the blood serum of the horse. *Biochem. J.*, 26, 2056-2064.

[37] Nachmansohn D., Machado A.L. (1943). The formation of acetylcholine. A new enzyme choline acetylase. *J. Neurophysiol.*, 6, 397-403.

[38] Mautner H.G. (1986). Choline acetyltransferase. In: *Neuromethods*, series 1, *Neurochemistry, Neurotransmitters Enzymes*. A.A. Boulton, G.B. Baker, and P.H. Yu (eds). Humana Press, Clifton, pp 273-317.

[39] Wessler I., Kirkpatrick C.J., Racke K. (1999). The cholinergic 'pitfall': acetylcholine, a universal cell molecule in biological systems, including humans. *Clin. Exp. Pharmacol. Physiol.*, 26, 198-205.

[40] Whittaker V.P. (1963). Identification of acetylcholine and related esters of biological origin. In: *Handbuch Der Experimentellen Pharmakologie, BD 15*. O. Eichler, A. Farah, G.B. Koelle (eds). Springer-Verlag, Berlin, pp 1-39.

[41] Stephenson M., Rowatt E. (1947). The production of acetylcholine by a strain of *Lactobacillus plantarum*. *J. Gen. Microbiol.*, 1, 279-298.

[42] Beyer G., Wense U.T. (1936). Über den Nachweis von Hormonen in einzelligen Tieren. Cholin und Acetylcholin in *Paramecium*. *Pflügers Arch. Gesamte Physiol. Menschen Tier.*, 237, 417-422.

[43] Lentz T.L. (1966). Histochemical localization of neurohumors in a sponge. *J. Exp. Zool.*, 162, 171-180.

[44] Erzen I., Brzin M. (1979). Cholinergic mechanisms in *Planaria torva*. *Comp. Biochem. Physiol.*, 64C, 207-216.

[45] Hartmann E., Kilbinger H. (1974). Occurrence of light-dependent acetylcholine concentrations in higher plants. *Experientia*, 30, 1387-1388.

[46] Grando S.A., Horton R.M. (1997). The keratinocyte cholinergic system: acetylcholine as an epidermal cytotransmitter. *Curr. Opin. Dermatol.*, 4, 262-268.

[47] Soreq H., Seidman S. (2001). Acetilcholinesterase - new roles for an old actor. *Nature Reviews. Neuroscience*, 2, 294-302.

[48] Buznikov G.A. (1990). Neurotransmitters in embryogenesis. In: *Physiology and General Biology*. Vol. 1. T.M. Turpaev (ed.). Harwood Academic Publishers, London, pp 1-526.

[49] Falugi C. (1993). Localization and possible role of molecules associated with the cholinergic system during "non-nervous" developmental events. *Eur. J. Histochem.*, 37, 287-294.

[50] Drews U. (1975). Cholinesterase in embryonic development. *Prog. Histochem. Cytochem.*, 7, 1-52.

[51] Falugi C., Prestipino G. (1987). Effects of some inhibitors of the cholinergic system during the sea urchin *Paracentrotus lividus* development. In: *Colloque International sur Paracentrotus lividus*. C.F. Boudoresque (ed.). Gis Posidonie, Marseille, pp 147-155.

[52] Buznikov G.A., Shmukler Y.B. (1981). Possible role of "prenervous" neurotransmitters in cellular interactions of early embryogenesis: a hypothesis. *Neurochem. Res.*, 6, 55-68.

[53] Minganti A., Falugi C., Raineri M., Pestarino M. (1981). Acetylcholinesterase in the embryonic development: an invitation to a hypothesis. *Acta Embryol. Morphol. Exp. n.s. 2,* 30-31.

[54] Stroud R.M., McCarthy M.P., Shuster M. (1990). Nicotinic acetylcholine receptor superfamily of ligand-gated ion channels. *Biochemistry*, 29, 11009-11023.

[55] Birdsall N.J.M., Burgen A.S.V., Hulme E.C. (1978). The binding of agonists to brain muscarinic receptors. *Mol. Pharmacol.*, 14, 723-736.

[56] Watson S., Arkinstall S. (1994). *The G-protein Linked Receptor Facts Book*. Academic Press, London.

[57] Mendel B., Rudney Y. (1943). Studies on Cholinesterase. 1. Cholinesterase and pseudo-cholinesterase. *Biochem. J.,* 37, 59-63.

[58] Talesa V., Contenti S., Mangiabene C., Pascolini R., Rosi G., Principato G.B. (1990). Propionylcholinesterase from *Murex brandaris*: comparison with other invertebrate cholinesterases. *Comp. Biochem. Physiol.*, 96C, 39-43.

[59] Massoulié J., Pezzementi L., Bon S., Krejici E., Valette F.M. (1993). Molecular and cellular biology of cholinesterases. *Prog. Neurobiol.*, 41, 41-91.

[60] Talesa V., Principato G.B., Mangiabene C., Giovannini E., Norton S.J., Rosi G. (1992). Cholinesterase in the cnidarians *Velella velella* (Hydrozoa: Syphonophora) and *Actinia equina* (Anthozoa: Actiniaria): a comparative study. *J. Exp. Zool.*, 263, 367-373.

[61] Augustinsson K.B. (1960). Butyryl- and propionylcholinesterases and related types of serine-sensitive esterases. In: *Enzymes*. Vol. IV. P. Boyer, H. Lardy, K. Myrback (eds). Academic Press, New York, pp 521-544.

[62] Hull R.W. (1961). Studies on suctorian protozoa: the mechanism of prey adherence. *J. Protozool.*, 8, 343-350.

[63] Doughty M.J., Dryl S. (1981). Control of ciliary activity in *Paramecium*: an analysis of chemosensory transduction in an eukaryotic unicellular organism. *Prog. Neurobiol.*, 15, 1-115.

[64] Verni F., Rosati G. (1986). Cytochemical demonstration of different esterases during the life cycle of the ciliate protozoon *Euplotes crassus*. *Arch. Protistenkd.*, 131, 95-99.

[65] Trielli F., Politi H., Falugi C., Delmonte Corrado M.U. (1997). Presence of molecules related to the cholinergic system in *Paramecium primaurelia* (Protista, Ciliophora) and possible role in mating pair formation: an experimental study. *J. Exp. Zool.*, 279, 633-638.

[66] Delmonte Corrado M.U., Politi H., Trielli F., Angelini C., Falugi C. (1999). Evidence for the presence of a mammalian-like cholinesterase in *Paramecium primaurelia* (Protista, Ciliophora) developmental cyle. *J. Exp. Zool.*, 283, 102-105.

[67] Delmonte Corrado M.U., Politi H., Ognibene M., Angelini C., Trielli F., Ballarini P., Falugi C. (2001). Synthesis of the signal molecole acetylcholine during the developmental cycle of *Paramecium primaurelia* (Protista, Ciliophora) and its possible function in conjugation. *J. Exp. Biol.*, 204, 1901-1907.

[68] Falugi C., Amaroli A., Evangelisti V., Viarengo A., Delmonte Corrado M.U. (2002). Cholinesterase activity and effects of its inhibition by neurotoxic drugs in *Dictyostelium discoideum*. *Chemosphere*, 48, 407-414.

[69] Amaroli A., Gallus L., Passalacqua M., Falugi C., Viarengo A., Delmonte Corrado M.U. (2003). Detection of cholinesterase activities and acetylcholine receptors during the developmental cycle of *Dictyostelium discoideum*. *Europ. J. Protistol.*, 39, 213-222.

[70] Delmonte Corrado M.U., Ognibene M., Trielli F., Politi H., Passalacqua M., Falugi C. (2002). Detection of molecules related to the GABAergic system in a single-cell eukaryote, *Paramecium primaurelia*. *Neurosci. Letters*, 329, 65-68.

[71] McMahon D. (1974). Chemical messengers in development: a hypothesis. *Science*, 185, 1012-1021.

[72] Rubino S., Mann S.K.O., Hori R.T., Pinko C., Firtel R.A. (1989). Molecular analysis of a developmentally regulated gene required for *Dictyostelium* aggregation. *Develop. Biol.*, 131, 27-36.

[73] Loomis W.F. (1975). *Dictyostelium discoideum. A developmental system*. Academic Press, New York.

[74] Escalante R., Vicente J.J. (2000). *Dictyostelium discoideum*: a model system for differentiation and patterning. *Int. J. Dev. Biol.*, 44, 819-835.

[75] Karnovsky M.J., Roots L. (1964). A simple thiocholine method for cholinesterases. *J. Histochem. Cytochem.*, 12, 219-221.

[76] Ellman G.L., Curtney K.D., Andres V., Featherstone R.M. (1961). A new and rapid colorimetric determination of acetylcholinesterase activity. *Biochem. Pharm.*, 7, 88-95.

[77] Oser B.L. (1965). *Hawk's Physiological Chemistry*. McGrow-Hill Book, New York.

[78] Ornstein N.L., Davis B. (1962). *Disk Electrophoresis*. Parts I and II. Distillation Products Industries, New York.

[79] Cartaud J., Bons S., Massoulié J. (1978). *Electrophorus* acetylcholinesterase. Biochemical and electron microscope characterization of low ionic strength aggregates. *J. Cell Biol.*, 77, 315-322.

[80] Dale H.H. (1914). The action of certain esters of choline and their relation to muscarine. *J. Pharmac. Exp. Ther.*, 6, 147-190.

[81] Hiwatashi K., Kitamura A. (1985). Fertilization in *Paramecium*. In: *Biology of Fertilization*. C.B. Metz, A. Monroy (eds). 2nd ed. Vol. 1. Academic Press, New York, pp 57-85.

[82] Miyake A. (1996). Fertilization and sexuality in Ciliates. In: *Ciliates - Cells as Organisms*. K. Hausmann, P.C. Bradbury (eds). Gustav Fischer, Stuttgart, pp 243-290.

[83] Sonneborn T.M. (1954). The relation of autogamy to senescence and rejuvenescence in *Paramecium aurelia*. *J. Protozool.*, 1, 38-53.

[84] Brandon C., Wu J.Y. (1978). Purification and properties of choline acetyltransferase from *Torpedo californica*. *J. Neurochem.*, 30, 791-797.

[85] Driskell W.J., Weber B.H., Roberts E. (1978). Purification of choline acetyltransferase from *Drosophila melanogaster*. *J. Neurochem.*, 30, 1135-1141.

[86] Malthe-Sørenssen D., Lea T., Fonnum F., Eskeland T. (1978). Molecular characterization of choline acetyltransferase from bovine brain caudate nucleus and some immunological properties of the highly purified enzymes. *J. Neurochem.*, 30, 35-46.

[87] Falugi C., Pieroni M., Moretti E. (1993). Cholinergic molecules and sperm function. *J. Submicrosc. Cytol. Pathol.*, 23, 63-69.

[88] Rattner B.A., Fairbrother A. (1991). Biological variability and the influence of stress on cholinesterase activity: In: *Cholinesterase-inhibition Insecticides: their Impact on Wildlife and the Environment. Chemicals in Agricolture.* Vol. 2. P. Mineau (ed.). Elsevier Science Publishers, Amsterdam, pp 89-108.

[89] Amaroli A., Trielli F., Falugi C., Corrado M.U. (2005). Valutazione tossicologica della presenza di fitofarmaci neurotossici nelle acque, utilizzando *Dictyostelium discoideum* (Protisti). *Notiziario dei Metodi Analitici IRSA/CNR*, in press.

[90] Lai H., Carino M. (1999). 60Hz magnetic field and central cholinergic activity: effects of exposure intensity and duration. *Bioelectromagnetics*, 20, 284-289.

[91] Amaroli A., Trielli F., Bianco B., Giordano S., Moggia E., Delmonte Corrado M.U. (2003). Time-varying ELF fields effects on a single-cell *Dictyostelium discoideum*. *6th International Congress of the European Bioelectromagnetics Association*, Budapest, 105.

[92] Soininen H., Unni L., Shillcutt S. (1990). Effect of acute and chronic cholinesterase inhibition on biogenic amines in rat brain. *Neurochem. Res.*, 15, 1185-1190.

INDEX

A

B

C

D

E

F

G

H

I

K

L

M

N

O

P

Q

R

S

W

X